Springer Series in Statistics

Advisors:
D. Brillinger, S. Fienberg, J. Gani,
J. Hartigan, K. Krickeberg

Springer Series in Statistics

Andrews/Herzberg: Data: A Collection of Problems from Many Fields for the Student and Research Worker.
Anscombe: Computing in Statistical Science through APL.
Berger: Statistical Decision Theory and Bayesian Analysis, 2nd edition.
Brémaud: Point Processes and Queues: Martingale Dynamics.
Brockwell/Davis: Time Series: Theory and Methods.
Daley/Vere-Jones: An Introduction to the Theory of Point Processes.
Dzhaparidze: Parameter Estimation and Hypothesis Testing in Spectral Analysis of Stationary Time Series.
Farrell: Multivariate Calculation.
Goodman/Kruskal: Measures of Association for Cross Classifications.
Hartigan: Bayes Theory.
Heyer: Theory of Statistical Experiments.
Jolliffe: Principal Component Analysis.
Kres: Statistical Tables for Multivariate Analysis.
Leadbetter/Lindgren/Rootzén: Extremes and Related Properties of Random Sequences and Processes.
LeCam: Asymptotic Methods in Statistical Decision Theory.
Manoukian: Modern Concepts and Theorems of Mathematical Statistics.
Miller, Jr.: Simultaneous Statistical Inference, 2nd edition.
Mosteller/Wallace: Applied Bayesian and Classical Inference: The case of *The Federalist Papers*.
Pollard: Convergence of Stochastic Processes.
Pratt/Gibbons: Concepts of Nonparametric Theory.
Read/Cressie: Goodness-of-Fit Statistics for Discrete Multivariate Data.
Sachs: Applied Statistics: A Handbook of Techniques, 2nd edition.
Senëta: Non-Negative Matrices and Markov Chains.
Siegmund: Sequential Analysis: Tests and Confidence Intervals.
Vapnik: Estimation of Dependences Based on Empirical Data.
Wolter: Introduction to Variance Estimation.
Yaglom: Correlation Theory of Stationary and Related Random Functions I: Basic Results.
Yaglom: Correlation Theory of Stationary and Related Random Functions II: Supplementary Notes and References.

Timothy R.C. Read
Noel A.C. Cressie

Goodness-of-Fit Statistics for Discrete Multivariate Data

Springer-Verlag
New York Berlin Heidelberg
London Paris Tokyo

Timothy R.C. Read
Hewlett-Packard Company
Palo Alto, CA 94304
USA

Noel A.C. Cressie
Department of Statistics
Iowa State University
Ames, IA 50011
USA

With 7 Illustrations

Mathematics Subject Classification (1980): 62H15, 62H17, 62E15, 62E20

Library of Congress Cataloging-in-Publication Data
Read, Timothy R.C.
Goodness-of-fit statistics for discrete multivariate data.
(Springer series in statistics)
Bibliography: p.
Includes indexes.
1. Goodness-of-fit tests. 2. Multivariate analysis.
I. Cressie, Noel A.C. II. Title. III. Series.
QA277.R43 1988 519.5′35 88-4648

Typeset by Asco Trade Typesetting Ltd., Hong Kong.
Printed and bound by R.R. Donnelley & Sons, Harrisonburg, Virginia.
Printed in the United States of America.

9 8 7 6 5 4 3 2 1

ISBN 0-387-96682-X Springer-Verlag New York Berlin Heidelberg
ISBN 3-540-96682-X Springer-Verlag Berlin Heidelberg New York

To Sherry for her support and encouragement (TRCR)
and
to Ray and Rene for giving me the education
they were unable to have (NACC)

Preface

The statistical analysis of discrete multivariate data has received a great deal of attention in the statistics literature over the past two decades. The development of appropriate models is the common theme of books such as Cox (1970), Haberman (1974, 1978, 1979), Bishop et al. (1975), Gokhale and Kullback (1978), Upton (1978), Fienberg (1980), Plackett (1981), Agresti (1984), Goodman (1984), and Freeman (1987).

The objective of our book differs from those listed above. Rather than concentrating on model building, our intention is to describe and assess the goodness-of-fit statistics used in the model verification part of the inference process. Those books that emphasize model development tend to assume that the model can be tested with one of the traditional goodness-of-fit tests (e.g., Pearson's X^2 or the loglikelihood ratio G^2) using a chi-squared critical value. However, it is well known that this can give a poor approximation in many circumstances.

This book provides the reader with a unified analysis of the traditional goodness-of-fit tests, describing their behavior and relative merits as well as introducing some new test statistics. The power-divergence family of statistics (Cressie and Read, 1984) is used to link the traditional test statistics through a single real-valued parameter, and provides a way to consolidate and extend the current fragmented literature. As a by-product of our analysis, a new statistic emerges "between" Pearson's X^2 and the loglikelihood ratio G^2 that has some valuable properties.

For completeness, Chapters 2 and 3 introduce the general notation and framework for modeling and testing discrete multivariate data that is used throughout the rest of the book. For readers totally unfamiliar with this field, we include many references to basic works which expand and supplement our discussion. Even readers who are familiar with loglinear models and the traditional goodness-of-fit statistics will find a new perspective in these chapters, together with a variety of examples.

The development and analysis of the power-divergence family presented here is based on consolidating and updating the results of our research papers (Cressie and Read, 1984; Read, 1984a, 1984b). The results are presented in a less terse style which includes worked examples and proofs of major results (brought together in the Appendix). We have included a literature survey on the most famous members of our family, namely Pearson's X^2 and the loglikelihood ratio statistic G^2, in the Historical Perspective. There are also a number of new results which have not appeared previously. These include a unified treatment of the minimum chi-squared, weighted least squares (minimum modified chi-squared), and maximum likelihood approaches to parameter estimation (Section 3.4); a detailed analysis of the contribution made by individual cells to the power-divergence statistic (Chapter 6), including Poisson-distributed cell frequencies with small expected values (Section 6.5); and a geometric interpretation of the power-divergence statistic (Section 6.6). In addition, Chapter 8 proposes some new directions for future research.

We consider Pearson's X^2 to be a sufficiently important statistic (omnipresent in all disciplines where quantitative data are analyzed) that it warrants an accessible and detailed reappraisal from our new perspective. Consequently this book is aimed at applied statisticians and postgraduate students in statistics as well as statistically sophisticated researchers and students in other disciplines, such as the social and biological sciences. With this audience in mind, we have emphasized the interpretation and application of goodness-of-fit statistics for discrete multivariate data. The technical discussion in each chapter is limited to that necessary for drawing comparisons between these goodness-of-fit statistics; detailed proofs are deferred to the Appendix.

Both of us have benefited from attending courses by John Darroch, from which the ideas of this research grew. We are very grateful for the detailed criticisms of Karen Kafadar, Ken Koehler, and an anonymous reviewer, which helped improve the presentation. Finally, Steve Fienberg's much appreciated comments led to improvements on the overall structure of the book.

Palo Alto, CA — Timothy R.C. Read
Ames, IA — Noel A.C. Cressie

Contents

CHAPTER 1

Introduction to the Power-Divergence Statistic

1.1. A Unified Approach to Model Testing

The definition and testing of models for discrete multivariate data has been the subject of much statistical research over the past twenty years. The widespread tendency to group data and to report group frequencies has led to many diverse applications throughout the sciences: for example, the reporting of survey responses (always, sometimes, never); the accumulation of patient treatment-response records (mild, moderate, severe, remission); the reporting of warranty failures (mechanical, electrical, no trouble found); and the collection of tolerances on a machined part (within specification, out of specification).

The statistical analysis of discrete multivariate data is the common theme of books such as Cox (1970), Haberman (1974, 1978, 1979), Bishop et al. (1975), Gokhale and Kullback (1978), Upton (1978), Fienberg (1980), Plackett (1981), Agresti (1984), Goodman (1984), and Freeman (1987). In these books the emphasis is on model development; it is usually assumed that the adequacy of the model can be tested with one of the traditional goodness-of-fit tests (e.g., Pearson's X^2 or the loglikelihood ratio G^2) using a chi-squared critical value. However this can be a poor approximation. In the subsequent chapters of this book, it is our intention to address this model-testing part of the inference process.

Our objective is to provide a unified analysis, depicting the behavior and relative merits of the traditional goodness-of-fit tests, by using the *power-divergence family of statistics* (Cressie and Read, 1984). The particular issues considered here include: the calculation of large-sample and small-sample significance levels for the test statistics; the comparison of large-sample and small-sample efficiency for detecting various types of alternative models;

sensitivity of the statistics to individual cell frequencies; and the effects of changing the cell boundaries.

Many articles have been published on goodness-of-fit statistics for discrete multivariate data; however this literature is widely dispersed and lacks cohesiveness. The power-divergence family of statistics provides an innovative way to unify and extend the current literature by linking the traditional test statistics through a single real-valued parameter.

1.2. The Power-Divergence Statistic

We shall present the basic form of the power-divergence statistic and provide a glimpse of some results to be developed in later chapters, by considering a simple example.

Suppose we randomly sample n individuals from a population (e.g., 100 women working for a large corporation). For each individual we measure a specific characteristic which must have only one of k possible outcomes (e.g., job classification: executive, manager, technical staff, clerical, other). Given a model for the behavior of the population (e.g., the distribution of women across the five job classifications is similar to that already measured for male employees, or measured for female employees five years previously), we can calculate how many individuals we *expect* for each outcome. The fit of the model can be assessed by comparing these *expected* frequencies for each outcome with the *observed* frequencies from our sample.

In this book we shall compare the observed and expected frequencies by using the *power-divergence statistic*

$$\frac{2}{\lambda(\lambda+1)}\sum_{i=1}^{k}\text{observed}_i\left[\left(\frac{\text{observed}_i}{\text{expected}_i}\right)^{\lambda}-1\right]; \tag{1.1}$$

where λ is a real-valued parameter that is chosen by the user. The cases $\lambda = 0$ and $\lambda = -1$ are defined as the limits $\lambda \to 0$ and $\lambda \to -1$, respectively. The power-divergence *family* consists of the statistics (1.1) evaluated for all choices of λ in the interval, $-\infty < \lambda < \infty$. Cressie and Read (1988) summarize the basic properties of the power-divergence statistic (1.1).

When the observed and expected frequencies match exactly for each possible outcome, the power-divergence statistic (1.1) is zero (for any choice of λ). In all other cases the statistic is positive and becomes larger as the observed and expected frequencies "diverge." Chapters 4 and 5 quantify how large (1.1) should be before we conclude that the model does not fit the data. Chapter 6 discusses the types of "divergence" measured best by different choices of λ.

Two very important special cases of the power-divergence statistic are Pearson's X^2 statistic (put $\lambda = 1$)

$$\sum_{i=1}^{k}\frac{(\text{observed}_i-\text{expected}_i)^2}{\text{expected}_i}; \tag{1.2}$$

and the loglikelihood ratio statistic G^2 (the limit as $\lambda \to 0$)

$$2 \sum_{i=1}^{k} \text{observed}_i \log\left[\frac{\text{observed}_i}{\text{expected}_i}\right]. \tag{1.3}$$

Other special cases are described in Section 2.4.

The power-divergence statistic (1.1) provides an important link between the well-known statistics (1.2) and (1.3). This link provides a mechanism to derive more general results about the behavior of these statistics in both large and small samples (Chapters 4–8). As a result of this analysis, the power-divergence statistic with $\lambda = 2/3$ emerges; i.e.,

$$\frac{9}{5} \sum_{i=1}^{k} \text{observed}_i \left[\left(\frac{\text{observed}_i}{\text{expected}_i}\right)^{2/3} - 1\right].$$

This statistic lies "between" X^2 and G^2 in terms of the parameter λ, and has some excellent properties (Chapter 5). These properties are explained partially by examining the role of the power transformations in Sections 6.5 and 6.6.

1.3. Outline of the Chapters

The major objective of this book is to provide a unified analysis and comparison of goodness-of-fit statistics for analyzing discrete multivariate data.

Chapter 2 introduces the general notation and framework for modeling and testing discrete data. The power-divergence statistic is defined more formally, and applied to an example which illustrates the weaknesses of the traditional approach to model testing.

Chapter 3 contains a general introduction to loglinear models for contingency tables. This is one of the most common applications for goodness-of-fit tests, and it motivates some of the results in the following chapters. This chapter includes a section on general methods of parameter estimation, model generation, and the subsequent selection of an appropriate model.

In Chapter 4, the emphasis shifts from defining models to testing the model fit. The asymptotic (i.e., large-sample) distribution of the power-divergence statistic is analyzed under both the *classical* (*fixed-cells*) assumptions (Sections 4.1 and 4.2) as well as *sparseness* assumptions (Section 4.3), which yield quite different results. The relative efficiency of power-divergence family members is also defined and compared for both local and nonlocal alternative models.

Chapter 5 discusses the appropriateness of using the asymptotic distributions (derived in Chapter 4) for small samples. Consequently some corrections to the asymptotic distributions are recommended on the basis of some exact calculations. Finally some exact power calculations are tabulated and compared with the asymptotic results of Chapter 4.

The sensitivity of the test statistics to individual cell contributions is examined and illustrated in Chapter 6. These results show how each member of the power-divergence family measures deviations from the model, which gives important new insights into the results of Chapter 5. Finally the moments of individual cell contributions are studied for cells with small expected frequencies, and a geometric interpretation is proposed to explain the good approximation of the power-divergence statistic with $\lambda = 2/3$ (observed in Chapter 5) to the asymptotic results of Chapter 4.

Throughout Chapters 3–6, the members of the power-divergence family are compared. Recommendations are made at the end of each chapter as to which statistic should be used for various models.

In Chapter 7 the literature on goodness-of-fit test statistics for *continuous* data is linked to research on the power-divergence statistic. In addition, the loss of information due to converting a continuous model to an observable discrete approximation (i.e., grouping the data) is discussed. The last section of this chapter is devoted to linking the literature on diversity indices and divergence measures from information theory to the current study of the power-divergence statistic.

Areas where we believe future research would be rewarding are presented in Chapter 8. These areas include methods of parameter estimation; finite sample approximations to asymptotic distributions; methods for choosing the appropriate "additive" scale for model fitting and analysis (including graphical analysis); methods to ensure selection of parsimonious models (Akaike's criterion); and general loss functions.

The Historical Perspective provides a detailed account of the literature for Pearson's X^2 and the loglikelihood ratio statistic G^2 over the last thirty years. We hope that this survey will be a useful resource for researchers working with traditional goodness-of-fit statistics for discrete multivariate data.

Finally the Appendix consists of a collection of proofs for results that are used throughout this book. Some of these proofs have appeared in the literature previously, but we have brought them together here to complete our discussion and make this text a more complete reference.

CHAPTER 2

Defining and Testing Models: Concepts and Examples

This chapter introduces the general notation and framework for modeling and testing discrete multivariate data. Through a series of examples we introduce the concept of a null model for the parameters of the sampling distribution in Section 2.1, and motivate the use of goodness-of-fit test statistics to check the null model in Section 2.2. A detailed example is given in Section 2.3, which illustrates how the magnitudes of the cell frequencies cause the traditional goodness-of-fit statistics to behave differently. An explanation of these differences is provided by studying the power-divergence statistic (introduced in Section 2.4) and is developed throughout this book. Section 2.5 considers an application from the area of visual perception.

2.1. Modeling Discrete Multivariate Data

Discrete multivariate data can arise in many contexts. In this book we shall draw on examples from a range of physical and social sciences. Our intention is to demonstrate the breadth of applications, rather than to give solutions to the important discrete multivariate problems in any one discipline. We shall start with an example taken from the medical sciences: patients with duodenal ulcers (Grizzle et al., 1969). The surgical operation for this condition usually involves some combination of vagotomy (i.e., cutting of the vagus nerve to reduce the flow of gastric juice) and gastrectomy (i.e., removal of part or all of the stomach). After such operations, patients sometimes suffer from a side effect called *dumping* or *postgastrectomy* syndrome. This side effect occurs after eating, due to an increased transit time of food, and is characterized by flushing, sweating, dizziness, weakness, and collapse of the vascular and nervous system response.

Suppose we define

$$Y_j = \begin{cases} 1 & \text{if patient } j \text{ exhibits this side effect} \\ 0 & \text{otherwise.} \end{cases} \tag{2.1}$$

Prior to collecting the data, Y_j is a *discrete random variable*: *discrete* because the possible values of Y_j are the integers $\{0, 1\}$, and *random* because prior to observing the patient response we do not know which value Y_j will take.

Discrete random variables may take more than two values (i.e., polytomous variables), as the following examples illustrate. Consider

(a) the response to a telephone survey regarding whether the respondent drives an American, European, Japanese, or other make of automobile (nonordered polytomous variable);
(b) the surface quality of raw laminate used to manufacture printed circuit boards measured as high quality, acceptable quality, or rejectable (ordered polytomous variable);
(c) the number of α particles emitted from a radioactive source over 30 seconds; possible values are $\{0, 1, 2, 3, \ldots\}$ (all nonnegative integers);
(d) the height of an individual, grouped into the categories short, medium, and tall (continuous variable grouped into discrete classes or cells).

The methods discussed in this book are applicable to all these types of discrete random variables, as will be illustrated in Chapter 3.

In the case of duodenal ulcer patients with random variables $\{Y_j\}$ defined by (2.1), the prevalence of dumping syndrome within a group of n patients is measured by

$$\begin{aligned} X_s &= \#\text{ patients (out of a group of } n\text{) exhibiting the side effect} \\ &= \sum_{j=1}^{n} Y_j. \end{aligned} \tag{2.2}$$

Here X_s is also a random variable, with possible values given by the set $\{0, 1, 2, \ldots, n\}$. For a group of $n = 96$ patients who underwent a drainage and vagotomy, Grizzle et al. (1969) report a total of $X_s = 35$ patients exhibiting symptoms of dumping syndrome.

If we assume each patient has an *equal* and *independent* chance π_s ($0 < \pi_s < 1$) of exhibiting this side effect, then the random variables $\{Y_j\}$ are independent and $Pr(Y_j = 1) = \pi_s$; $j = 1, \ldots, n$. Frequently an experimenter will wish to check some preconceived hypothesis regarding the value of the probability π_s. For example, prior to collecting any data, it may have been hypothesized that the incidence of dumping severity for patients undergoing drainage and vagotomy is similar to that for postvagotomy diarrhea, which is generally around 25%. We write this hypothesis as

$$H_0\colon \pi_s = 0.25. \tag{2.3}$$

Therefore for any group of $n = 96$ patients, we would expect (on average)

$n\pi_s = 96 \times 0.25 = 24$ patients in the group to exhibit the symptoms of dumping syndrome.

Alternatively, the experimenter may wish to hypothesize that the incidence of dumping syndrome for patients who undergo a hemigastrectomy (i.e., removal of half the stomach) and vagotomy is the same as that for patients who undergo the less severe operation of drainage and vagotomy. In this case, we assume neither incidence is known in advance and they must be obtained from observing patients. This hypothesis is considered in detail in Section 3.2 where a more comprehensive description of the data from Grizzle et al. (1969) is presented.

This example illustrates two important concepts that are central to this book. First is the concept of the *sampling distribution*: In this case, we are assuming the random variables Y_j; $j = 1, \ldots, n$ are independent and $Pr(Y_j = 1) = \pi_s$ for each j. Consequently, X_s in (2.2) has a *binomial distribution* with parameters n and π_s, described in more detail later. Second is the concept of a hypothesized *null model* for the parameter(s) of the sampling distribution, illustrated by equation (2.3).

The choice of an appropriate sampling distribution, and the subsequent selection of an appropriate null model, are very important tasks. Chapter 3 is devoted to discussing these issues. However the focus of the subsequent chapters of this book is on the description and comparison of statistics used to *evaluate*, or *test*, how well the null model describes a given data set.

More generally, consider observing a random variable Y, which can have one of k possible outcomes; $c_1, c_2, \ldots, c_k$ with probabilities $\pi_1, \pi_2, \ldots, \pi_k$. We assume that the outcome cells (or categories) $c_1, c_2, \ldots, c_k$ are mutually exclusive and $\sum_{i=1}^{k} \pi_i = 1$. For example, Y could represent the answer to the question "Do you wear a seatbelt?" with $k = 3$ possible responses; $c_1 =$ "always," $c_2 =$ "sometimes," and $c_3 =$ "never." If n realizations of Y are observed (e.g., n people are interviewed on seatbelt usage), we can summarize the responses with the random vector $\mathbf{X} = (X_1, X_2, \ldots, X_k)$ where

$$X_i = \# \text{ times (out of } n\text{) that } Y = c_i; \qquad i = 1, \ldots, k;$$

note that $\sum_{i=1}^{k} X_i = n$. Provided the observed Y's are *independent* and *identically distributed*, then $\mathbf{X}$ has a *multinomial distribution* with parameters n, $\boldsymbol{\pi} = (\pi_1, \pi_2, \ldots, \pi_k)$, which we write $\text{Mult}_k(n, \boldsymbol{\pi})$. The probability of any particular outcome $\mathbf{x} = (x_1, x_2, \ldots, x_k)$, is then

$$Pr(\mathbf{X} = \mathbf{x}) = n! \prod_{i=1}^{k} \frac{\pi_i^{x_i}}{x_i!}, \tag{2.4}$$

where $0 \le x_i \le n$, $0 \le \pi_i \le 1$; $i = 1, \ldots, k$, and $\sum_{i=1}^{k} x_i = n$, $\sum_{i=1}^{k} \pi_i = 1$. Throughout we use lowercase $\mathbf{x}$ to represent a realization of the uppercase random vector $\mathbf{X}$. When $k = 2$, the *multinomial distribution* (2.4) reduces to the *binomial distribution* described earlier for the dumping syndrome example.

Once the response variable has been reformulated as a multinomial random vector, the question of modeling the response reduces to developing a null model for the multinomial probability vector $\boldsymbol{\pi}$:

$$H_0: \boldsymbol{\pi} = \boldsymbol{\pi}_0 \tag{2.5}$$

where $\boldsymbol{\pi}_0 = (\pi_{01}, \pi_{02}, \ldots, \pi_{0k})$ is the hypothesized probability vector. From this cell probability vector $\boldsymbol{\pi}_0$, we obtain the *expected* cell frequencies (from the null model) to be $n\boldsymbol{\pi}_0 = (n\pi_{01}, n\pi_{02}, \ldots, n\pi_{0k})$. For the dumping syndrome example we have $k = 2$, with c_1 = "side effect," c_2 = "no side effect." The hypothesis in (2.3) thus becomes

$$H_0: \boldsymbol{\pi} = (\pi_1, \pi_2) = (\pi_s, 1 - \pi_s) = (0.25, 0.75),$$

with $\pi_1 + \pi_2 = \pi_s + (1 - \pi_s) = 1$. For $n = 96$, the *expected* frequency in cell c_1 is $96 \times 0.25 = 24$ patients, and in cell c_2 is $96 \times 0.75 = 72$ patients. From the results reported by Grizzle et al. (1969), the binomial random vector $\mathbf{X} = (X_1, X_2) = (X_s, 96 - X_s)$ has observed value $\mathbf{x} = (35, 61)$, indicating that 35 patients exhibited the side effect while 61 did not show any symptoms. We shall return to this example in the next section, where we test the fit of the hypothesis (2.3) to the observed data.

The multinomial sampling distribution is not the only possible distribution available for data grouped into k cells. In Chapter 3, the *product-multinomial* and *Poisson* sampling distributions are introduced as two other candidates that are often appropriate. Fortunately, most of the results on modeling under these three distributions turn out to be the same, so we consider only the multinomial distribution for illustrative purposes in this chapter.

In all the examples discussed so far, only one random variable has been observed on each sampled individual. Frequently we may observe many random variables on each individual. For example, consider the survey on seatbelt usage: In addition to the question "Do you wear a seatbelt?" a second question might be "Have you ever been involved in a car accident requiring subsequent hospitalization?" The *joint* answer to these two questions provides information on two random variables *simultaneously*, as illustrated in Table 2.1. The joint frequencies in Table 2.1 can be represented by a multinomial random vector $\mathbf{X} = (X_{11}, X_{12}, X_{13}, X_{21}, X_{22}, X_{23})$ and probability vector $\boldsymbol{\pi} = (\pi_{11}, \pi_{12}, \pi_{13}, \pi_{21}, \pi_{22}, \pi_{23})$ satisfying $\sum_{i=1}^{2}\sum_{j=1}^{3} X_{ij} = n$ and $\sum_{i=1}^{2}\sum_{j=1}^{3} \pi_{ij} = 1$. This example illustrates that the vectors $\mathbf{X}$ and $\boldsymbol{\pi}$

Table 2.1. Cross-Classified Response Table for Seatbelt Usage versus Previous Accident History

Previous accident requiring hospital	Use seatbelt: Never	Sometimes	Always
yes	X_{11}	X_{12}	X_{13}
no	X_{21}	X_{22}	X_{23}

may have substantial inner structure, but are represented as row vectors for ease of notation.

We refer to data sets involving two or more variables collected simultaneously as *multivariate data. Multivariate analysis* provides information about the *simultaneous relationships* between two or more variables. The current availability of fast computers, statistical software, and large accessible data bases has resulted in a substantial increase in the use of multivariate analysis techniques. Chapter 3 takes up the case of multivariate data in detail, and reviews the general methodology for developing models for $\boldsymbol{\pi}$ to measure the association between two or more variables. For example, in the seatbelt study we might want to develop a model for $\boldsymbol{\pi}$ that assumes independence of accident history and seatbelt usage.

The seatbelt example described earlier illustrates the simultaneous measurement and analysis of two *response* variables. In some cases, values are collected on *explanatory* variables as well as *response* variables. The term *explanatory* indicates the variable is *fixed* at certain levels by the experimenter. For example, the temperature and pH controls on a chemical bath can be controlled and recorded along with the response variable, "plating thickness." A second example would be the assignment of different drug therapies to a group of study patients and their subsequent response(s) relative to their assigned therapy. The methodology described in Chapter 3, and the subsequent results for goodness-of-fit tests presented in the remaining chapters, can be used for both *response* and *explanatory* variables.

Throughout this book, we shall concentrate on methods for *discrete* data. Methods for modeling and analyzing *continuous* multivariate data (e.g., length, time, weight), which have not been grouped into discrete categories, are covered well by such books as Cochran and Cox (1957), Draper and Smith (1981), Muirhead (1982), Anderson (1984), and Dillon and Goldstein (1984).

2.2. Testing the Fit of a Model

Suppose now that the null model (2.5) has been hypothesized and the data have been collected and summarized into a vector of frequencies $\mathbf{x} = (x_1, x_2, \ldots, x_k)$, where x_i represents the number of responses in cell c_i and $\sum_{i=1}^{k} x_i = n$. The fit of the model is usually assessed by comparing the frequencies *expected* in each cell, given by $n\boldsymbol{\pi}_0$, against the *observed* frequencies $\mathbf{x}$. If there is substantial discrepancy between the observed frequencies and those expected from the null model, then it would be wise to *reject* the null model, and look for some alternative that is more concordant with the data. *Goodness-of-fit tests* use the properties of a hypothesized distribution to assess whether data are generated from that distribution. In this book, we shall deal almost exclusively with tests based on how well expected frequencies fit observed frequencies.

The most well known goodness-of-fit statistics used to test (2.5) are Pearson's

X^2 (introduced by Pearson, 1900);

$$X^2 = \sum_{i=1}^{k} \frac{(x_i - n\pi_{0i})^2}{n\pi_{0i}}, \tag{2.6}$$

and the loglikelihood ratio statistic;

$$G^2 = 2 \sum_{i=1}^{k} x_i \log(x_i/n\pi_{0i}). \tag{2.7}$$

When there is perfect agreement between the null model and the data, $n\pi_{0i} = x_i$ for each $i = 1, \ldots, k$, and consequently $X^2 = 0 = G^2$. As the discrepancy between $\mathbf{x}$ and $n\boldsymbol{\pi}_0$ increases, so too do the values of X^2 and G^2, although in general the statistics are no longer equal.

If $\mathbf{x}$ in (2.6) and (2.7) is replaced by the multinomial random vector $\mathbf{X}$, then X^2 and G^2 are random variables. Assuming the null model (2.5), Pearson (1900) derives the asymptotic (i.e., as sample size n increases) distribution of X^2 to be a chi-squared distribution with $k - 1$ degrees of freedom. The same asymptotic distribution attains for G^2, a result which Wilks (1938) proves in a more general context. Subsequently it has been shown, under certain conditions on $\boldsymbol{\pi}_0$ and k, that X^2 and G^2 are asymptotically equivalent (Neyman, 1949); see Chapter 4.

For example, consider again the incidence of dumping syndrome for patients undergoing drainage and vagotomy as described in Section 2.1. We are now in a position to test the fit of the hypothesized incidence rate of 25% given by (2.3). Here, $k = 2$ with observed frequencies $\mathbf{x} = (x_1, x_2) = (35, 61)$ and expected frequencies $n\boldsymbol{\pi}_0 = (n\pi_{01}, n\pi_{02}) = (24, 72)$, which give $X^2 = 6.72$ and $G^2 = 6.18$. Both of these values far exceed the 95th percentile of a chi-squared distribution with one degree of freedom given by $\chi_1^2(0.05) = 3.84$. In other words, there is less than a 5% chance of observing values of X^2 (similarly for G^2) greater than 3.84, if indeed the hypothesized incidence rate of 25% is reasonable. On the basis of the observed values for X^2 and G^2, and the observed incidence rate of 35 patients out of 96, or 36%, we conclude that the 25% incidence rate model is too low to be supported by the data. We shall return to this example in more detail in Chapter 3.

A further complication arises if the null model does not completely specify the null probability vector $\boldsymbol{\pi}_0$. For example, in Section 2.3 we consider a model for $\boldsymbol{\pi}$ given by

$$\log(\pi_i) = \alpha + \beta i,$$

that is, a model linear in the logarithms of the probabilities with two unspecified parameters α and β. Chapter 3 considers more general loglinear models with unspecified parameters (generally called *nuisance* parameters) for multidimensional data. Regardless of the justification of the model, such nuisance parameters must be estimated from the sample before the expected frequencies and goodness-of-fit statistics (2.6) and (2.7) can be calculated.

More generally, we write the null model (2.5) as

$$H_0 \colon \boldsymbol{\pi} \in \Pi_0, \tag{2.8}$$

where Π_0 represents a specified *set* of probability vectors that are hypothesized for $\boldsymbol{\pi}$. Estimating nuisance parameters can be thought of as choosing a particular element of the set Π_0 which is "most consistent" with the data in the sample. We denote such an *estimated* probability vector by $\hat{\boldsymbol{\pi}}$.

Provided an *efficient* method is employed to estimate the nuisance parameters (i.e., to choose $\hat{\boldsymbol{\pi}} \in \Pi_0$), Fisher (1924) shows in the case of one nuisance parameter that

$$X^2 = \sum_{i=1}^{k} \frac{(X_i - n\hat{\pi}_i)^2}{n\hat{\pi}_i} \tag{2.9}$$

is asymptotically chi-squared with $k - 2$ degrees of freedom when (2.8) is true. Pearson (1900) had originally recommended maintaining $k - 1$ degrees of freedom in this situation. Subsequent generalizations to s parameters (e.g., Cramér, 1946) produced the following well-known result.

Assume the null model (2.8) is true, and certain regularity conditions on $\boldsymbol{\pi}$ and k hold, then both X^2 in (2.9) and

$$G^2 = 2 \sum_{i=1}^{k} X_i \log(X_i/n\hat{\pi}_i) \tag{2.10}$$

are asymptotically chi-squared with degrees of freedom reduced to $k - s - 1$. In other words, one degree of freedom is subtracted for each parameter estimated (efficiently).

Sections 3.4 and 4.1 provide a detailed discussion of what constitutes an *efficient* estimate, and Section 4.1 describes the regularity conditions required of $\boldsymbol{\pi}$ and k. Section 4.3 discusses some situations in which these regularity conditions do not hold, and shows that X^2 and G^2 are *no longer* asymptotically equivalent.

Various other goodness-of-fit statistics have been proposed over the last forty years. These include the Freeman-Tukey statistic (Freeman and Tukey, 1950; Bishop et al., 1975), which, following Fienberg (1979) and Moore (1986), we define as

$$F^2 = 4 \sum_{i=1}^{k} (\sqrt{X_i} - \sqrt{n\hat{\pi}_i})^2; \tag{2.11}$$

the modified loglikelihood ratio statistic or *minimum discrimination information* statistic for the *external constraints problem* (Kullback, 1959, 1985; see also Section 3.5)

$$GM^2 = 2 \sum_{i=1}^{k} n\hat{\pi}_i \log(n\hat{\pi}_i/X_i); \tag{2.12}$$

and the Neyman-modified X^2 statistic (Neyman, 1949)

$$NM^2 = \sum_{i=1}^{k} \frac{(X_i - n\hat{\pi}_i)^2}{X_i}. \tag{2.13}$$

All three of these statistics have been shown by various authors to have the same asymptotic chi-squared distribution as X^2 and G^2, under the conditions outlined earlier (see Section 4.1 for a detailed discussion of this result). However, for finite sample size these five statistics are *not* equivalent, and there has been much argument in the literature as to which statistic is the "best" to use. A detailed account of these results is contained in the Historical Perspective at the end of this book.

2.3. An Example: Time Passage and Memory Recall

Consider the possible relationship between our ability to recall specific events and the amount of time that has passed since the event has occurred. Haberman (1978, pp. 2–23) provides an example of such a data set, which originally formed part of a larger study into the relationship between life stresses and illnesses in Oakland, California. For the particular relationship considered here, each respondent was asked to note which stressful events, out of a list of 41, had occurred within the last 18 months. Table 2.2 gives the total number of respondents for each month who indicated one stressful event between 1 and 18 months prior to interview.

A quick perusal of this table indicates that the respondents' recall decreases with time. This observation is supported by testing the null hypothesis that the probability of dating an event in month i is equal for all i, that is,

$$H_0\colon \pi_i = 1/18; \qquad i = 1, \ldots, 18, \tag{2.14}$$

where $\pi_i = Pr$ (an individual dates an event in month i). For the sample size $n = 147$, this gives an expected frequency of $n\pi_i = 147/18 = 8.17$ respondents in each of the 18 cells (Table 2.2).

Taking the observed and expected frequencies from Columns 2 and 3 of Table 2.2, the calculated values of the goodness-of-fit statistics, given by (2.9)–(2.13), are $X^2 = 45.3$; $G^2 = 50.7$; $F^2 = 57.7$; $GM^2 = 70.8$; and $NM^2 = 136.6$. Recall from Section 2.2 that if the null hypothesis (2.14) is true, we expect each of these five values to be a realization from a distribution which is approximately chi-squared with $k - 1 = 18 - 1$ degrees of freedom. The 95th percentile of a chi-squared random variable (i.e., the 5% critical value) with seventeen degrees of freedom is $\chi^2_{17}(0.05) = 27.59$; consequently there is only a small chance (5%) of observing a value greater than or equal to 27.59 if indeed the hypothesized model for $\boldsymbol{\pi}$ is reasonable. All five test statistics have values far in excess of 27.59. We conclude the equiprobable hypothesis is untenable, regardless of which statistic we use.

In the light of this result, Haberman proposes a loglinear time-trend model as a more appropriate explanation of these data, that is,

$$H_0\colon \log(\pi_i) = \alpha + \beta i; \qquad i = 1, \ldots, 18, \tag{2.15}$$

for some unknown constants α and β. Putting $\alpha = \log(1/18)$ and $\beta = 0$, we obtain the equiprobable model (2.14) as a special case. For $\beta < 0$, the model

Table 2.2. Observed and Expected Frequencies by Month prior to Interview of Stressful Events Reported by Subjects

Months before interview	Observed frequency	Expected frequency from (2.14)	Expected frequency from (2.15)
1	15	8.17	15.171
2	11	8.17	13.952
3	14	8.17	12.831
4	17	8.17	11.800
5	5	8.17	10.852
6	11	8.17	9.980
7	10	8.17	9.178
8	4	8.17	8.440
9	8	8.17	7.762
10	10	8.17	7.138
11	7	8.17	6.565
12	9	8.17	6.037
13	11	8.17	5.552
14	3	8.17	5.106
15	6	8.17	4.696
16	1	8.17	4.318
17	1	8.17	3.971
18	4	8.17	3.652
Total	147	147.00	147.00

Subjects are limited to those reporting one stressful event between 1 and 18 months prior to interview.

Expected frequencies are based on the equiprobable model (2.14) and the loglinear time-trend model (2.15).

Source: Haberman (1978, pp. 2–23).

suggests that as the months i prior to interview *increase*, the probability π_i of dating an event in month i *decreases* at an exponential rate. Haberman (1978, pp. 7–9) contends the loglinear time-trend model is consistent with many models for memory recall and further motivates (2.15) by appealing to an exponential decay model for memory.

The constants α and β must be estimated from the sample before the expected values under (2.15) can be calculated. Haberman uses the method of maximum likelihood (described in detail in Section 3.4) to estimate α and β. The calculations require an iterative solution (Haberman, 1978, p. 10), which converges to give the estimated expected values in Table 2.2, Column 4.

We are now in a position to test the fit of the null model (2.15). However, the estimation of the two parameters α and β means that we may no longer use $18 - 1 = 17$ degrees of freedom for the reference chi-squared distribution, used in determining the significance of the calculated goodness-of-fit statistic. At first glance, it appears from the results of Section 2.2 that we should subtract two additional degrees of freedom from the seventeen, one for each parameter estimated. A more careful check of (2.15) shows that $\alpha = -\log(\sum_{i=1}^{k} e^{\beta i})$, and therefore there is really only one extra parameter estimated. Consequently we subtract only one further degree of freedom for the estimation of the expected values, giving sixteen degrees of freedom. The corresponding 95th percentile of the chi-squared distribution is $\chi^2_{16}(0.05) = 26.3$, against which we shall compare the calculated values of the goodness-of-fit statistics.

Haberman reports the values for Pearson's X^2 and the loglikelihood ratio statistics to be $X^2 = 22.7$ and $G^2 = 24.6$. Since both values are less than 26.3, there is a "reasonable" probability that they are realizations of a chi-squared random variable with sixteen degrees of freedom, from which Haberman (1978, p. 16) concludes "... the model is consistent with the data, although the fit is not perfect."

If we decide to use the Freeman-Tukey or modified loglikelihood ratio statistics, we obtain $F^2 = 26.5$ and $GM^2 = 29.5$. Both of these tests indicate we should reject the time-trend model at the approximate 5% significance level, since they are greater than 26.3. Even more alarming is the value of the Neyman-modified X^2 statistic $NM^2 = 40.6$, which is almost *twice* the value of X^2, and recommends a strong rejection of the time-trend model.

What causes such different results? More importantly, how can we decide whether to accept or reject the model? Answering these questions requires an understanding of: (a) the limitations in using the chi-squared distribution to approximate the exact critical value of the test, and (b) the type of departure from H_0 that the different statistics measure. Topic (a) is covered in detail in Chapter 5, and topic (b) is the subject of Chapter 6, where we shall return to this example.

The problem of differing test results in the memory-recall example is not an isolated pathological case. For example, Fienberg (1980, pp. 35–43) analyzes a multivariate data set describing the structural habitat of the Anolis lizards of Bimini. The habitats of two species of male lizards were observed, and for each the perch height and perch diameter were recorded. Using the loglinear approach described in Section 3.3, Fienberg focuses in on a model specifying conditional independence of perch height and perch diameter, given the species. This model fitted a previous data set quite well, however for the current data Fienberg (1980, p. 43) notes "... there is some question as to whether this conditional-independence model fits, since X^2 exceeds the 0.05 significance value while G^2 does not."

2.4. Applying the Power-Divergence Statistic

To understand the differences between the traditional statistics introduced in Section 2.2, we need to draw some comparisons between their behavior and characteristics. Cressie and Read (1984) introduce the *power-divergence family* of goodness-of-fit statistics, and show the statistics defined by (2.9)–(2.13) to be members of a single-parameter family (see also Moore, 1986). This linking provides a new perspective from which to compare these traditional statistics and indicates some new statistics that are highly competitive.

The *power-divergence statistic* is defined as

$$2nI^{\lambda}(\mathbf{X}/n:\hat{\boldsymbol{\pi}}) = \frac{2}{\lambda(\lambda+1)}\sum_{i=1}^{k} X_i\left[\left(\frac{X_i}{n\hat{\pi}_i}\right)^{\lambda} - 1\right]; \qquad -\infty < \lambda < \infty \tag{2.16}$$

where λ is the family parameter. The term *power divergence* describes the fact that the statistic $2nI^{\lambda}(\mathbf{X}/n:\hat{\boldsymbol{\pi}})$ measures the divergence of $\mathbf{X}/n$ from $\hat{\boldsymbol{\pi}}$ through a (weighted) sum of powers of the terms $X_i/n\hat{\pi}_i$ for $i = 1, \ldots, k$. Measures that determine the divergence of two probability distributions have a long history (Chapter 7). One such family of measures specifies the divergence of the probability distributions $\mathbf{p} = (p_1, p_2, \ldots, p_k)$ and $\mathbf{q} = (q_1, q_2, \ldots, q_k)$ to be

$$I^{\lambda}(\mathbf{p}:\mathbf{q}) = \frac{1}{\lambda(\lambda+1)}\sum_{i=1}^{k} p_i\left[\left(\frac{p_i}{q_i}\right)^{\lambda} - 1\right]; \qquad -\infty < \lambda < \infty.$$

In (2.16) we are measuring how far the empirical probability distribution $\mathbf{X}/n$ diverges from the hypothesized distribution $\hat{\boldsymbol{\pi}}$.

In those chapters where we are more concerned with comparing the cell frequency vector $\mathbf{X}$ against the expected frequency vector $\hat{\mathbf{m}} = n\hat{\boldsymbol{\pi}}$, we shall use the notation

$$2I^{\lambda}(\mathbf{X}:\hat{\mathbf{m}}) = \frac{2}{\lambda(\lambda+1)}\sum_{i=1}^{k} X_i\left[\left(\frac{X_i}{\hat{m}_i}\right)^{\lambda} - 1\right]; \qquad -\infty < \lambda < \infty, \tag{2.17}$$

for the power-divergence statistic, where $\sum_{i=1}^{k} X_i = \sum_{i=1}^{k} \hat{m}_i$. Notice that now the total sample size n no longer appears explicitly. From the context, it shall always be clear as to which form of the power-divergence statistic we are using. The rest of this chapter refers to the form $2nI^{\lambda}(\mathbf{X}/n:\hat{\boldsymbol{\pi}})$ in (2.16).

Equation (2.16) is undefined for $\lambda = -1$ or $\lambda = 0$. However, if we define these cases by the continuous limits of (2.16) as $\lambda \to -1$ and $\lambda \to 0$, then $2nI^{\lambda}(\mathbf{X}/n:\hat{\boldsymbol{\pi}})$ is continuous in λ. Using the fact that $\log(t) = \lim_{h\to 0}(t^h - 1)/h$, we obtain

$$\lim_{\lambda\to 0} 2nI^{\lambda}(\mathbf{X}/n:\hat{\boldsymbol{\pi}}) = 2\sum_{i=1}^{k} X_i \log(X_i/n\hat{\pi}_i)$$

and

$$\lim_{\lambda\to -1} 2nI^{\lambda}(\mathbf{X}/n:\hat{\boldsymbol{\pi}}) = 2\sum_{i=1}^{k} n\hat{\pi}_i \log(n\hat{\pi}_i/X_i).$$

All the statistics considered in Section 2.2 can now be expressed as members of the power-divergence family (2.16). From equations (2.10) and (2.12), we observe

$$2nI^{0}(\mathbf{X}/n:\hat{\boldsymbol{\pi}}) = G^2$$

and

$$2nI^{-1}(\mathbf{X}/n : \hat{\boldsymbol{\pi}}) = GM^2.$$

Furthermore, if we evaluate (2.16) with $\lambda = 1$, $-1/2$, and -2, a quick comparison with (2.9), (2.11), and (2.13) yields

$$2nI^{1}(\mathbf{X}/n : \hat{\boldsymbol{\pi}}) = X^2,$$

$$2nI^{-1/2}(\mathbf{X}/n : \hat{\boldsymbol{\pi}}) = F^2,$$

and

$$2nI^{-2}(\mathbf{X}/n : \hat{\boldsymbol{\pi}}) = NM^2.$$

Now let us reconsider the memory-recall example of Section 2.3. For the loglinear time-trend model (2.15), there was some question regarding the model fit, due to the discrepancies among the calculated goodness-of-fit statistics X^2, G^2, F^2, GM^2, and NM^2. Table 2.3 contains the calculated values of the power-divergence statistic for some values of λ in the range $[-10, 10]$, based on the time-trend model. If we use the chi-squared 95th percentile given by $\chi^2_{16}(0.05) = 26.3$, then we accept the loglinear time-trend model at the 5% significance level when $-1/2 < \lambda < 5$. However if λ is outside this range, we reject the model.

There is a clear relationship in Table 2.3 between the value $2nI^{\lambda}(\mathbf{x}/n : \hat{\boldsymbol{\pi}})$

Table 2.3. Values of the Power-Divergence Statistic (2.16) Calculated from Table 2.2 for the Time-Trend Model

λ	$2nI^{\lambda}(\mathbf{x}/n : \hat{\boldsymbol{\pi}})$
−10.00	72.2×10^3
−5.00	28.9×10
−3.00	65.6
−2.00	40.6
−1.50	34.0
−1.00	29.5
−0.50	26.5
0.00	24.6
0.50	23.4
0.67	23.1
1.00	22.7
1.50	22.6
2.00	22.9
3.00	24.8
5.00	35.5
10.00	21.4×10

and the parameter λ. As $|\lambda - 1.5|$ increases, so too does $2nI^{\lambda}(\mathbf{x}/n : \hat{\boldsymbol{\pi}})$. Why does this occur? Is it appropriate to use the $\chi^2_{16}(0.05)$ critical value for all values of λ? What is the nature of the departure from the null model that each statistic is measuring?

Our objective in this book is to answer these questions, and others that arise during the ensuing discussion. The power-divergence statistic offers a clearer perspective from which to unify and extend the current literature on goodness-of-fit tests, previously limited to looking *separately* at each test statistic.

2.5. Power-Divergence Measures in Visual Perception

Before proceeding with the discussion of goodness-of-fit tests we examine briefly another example, this one from the area of visual perception, that motivates the use of measures of divergence of the form (2.16).

Individuals are shown a two-dimensional map whose area is divided into different-colored contiguous regions $R_1, R_2, \ldots, R_k$ with areas $A_1, A_2, \ldots, A_k$, respectively. They are asked to guess the areas of the colors relative to say, the smallest colored region R_1. Stevens' Law (Stevens, 1975) says that the *perceived* areas relative to R_1 are $\{(A_i/A_1)^{\lambda}: i = 1, \ldots, k\}$, where λ typically lies in the range from 0.6 to 0.9. Cleveland (1985, pp. 243–245) has a general description of Stevens' Law as applied to lengths, areas, and volumes; he emphasizes that these empirically observed values of λ should be used to understand the perception biases that some graphical displays can produce.

Setting

$$p_i = A_i/(A_1 + A_2 + \cdots + A_k); \qquad i = 1, \ldots, k,$$

Stevens' Law says that the perceived relative importance of regions R_i and R_j is

$$(p_i/p_j)^{\lambda}.$$

Now take the original map, and enlarge and contract various of the colored regions $\{R_i: i = 1, \ldots, k\}$ to $\{R_i': i = 1, \ldots, k\}$ (using the same colors and ensuring the total area remains the same). Define A_i' and p_i' in an analogous way; notice that $\sum_{i=1}^{k} A_i = \sum_{i=1}^{k} A_i'$. Then the relative difference between the *perceived* areas of regions R_i' and R_i is

$$(p_i'/p_i)^{\lambda} - 1,$$

which, when squared, weighted proportional to area, and summed, yields the measure of difference

$$\sum_{i=1}^{k} p_i \left[\left(\frac{p_i'}{p_i} \right)^{\lambda} - 1 \right]^2.$$

From the results of Section 6.6 (equation (6.17)), this is approximately proportional to the power-divergence measure

$$I^{\lambda}(\mathbf{p}' : \mathbf{p}) = \frac{1}{\lambda(\lambda+1)} \sum_{i=1}^{k} p_i' \left[\left(\frac{p_i'}{p_i} \right)^{\lambda} - 1 \right].$$

The empirical evidence of Stevens (1975) suggests that λ ranges from 0.6 to 0.9, which incidentally contains $\lambda = 2/3$. The value $\lambda = 2/3$ defines an important member of the power-divergence family from various different perspectives, as we shall see in the following chapters. Notice that when $\lambda = 1$ (where perceived area = actual area), our analysis suggests that a Pearson X^2 type divergence is the most appropriate measure of map difference.

CHAPTER 3

Modeling Cross-Classified Categorical Data

In this chapter we introduce some of the basic terminology and concepts for fitting loglinear models to cross-classified categorical data. We discuss briefly those aspects of model fitting relevant to the development of further chapters, and we show how the power-divergence statistic adds a new dimension to categorical data analysis. In addition, Section 3.4 describes methods of estimating unknown model parameters from the perspective of the power-divergence statistic. In Section 3.5 we discuss the *minimum discrimination information* approach to characterizing the loglinear model, and illustrate how the power-divergence statistic provides a generalization that characterizes other models (including the linear model). The chapter concludes with a discussion in Section 3.6 of strategies for choosing an appropriate loglinear model.

3.1. Association Models and Contingency Tables

We have seen in Chapter 2 how goodness-of-fit tests are used to assess whether or not a set of data comes from some given discrete distribution or class of distributions. A second, and perhaps more frequent, application of goodness-of-fit tests is to draw inferences about *associations* between two or more random variables: Does knowledge of the value of one random variable offer any insight into the distribution of a second? For example, does knowledge of blood pressure (or serum cholesterol) tell us anything about a person's predisposition to coronary heart disease?

Whenever each member of a sample is classified simultaneously according to two or more variables, the resultant table of frequencies for each cross-classification is called a *contingency table*. For example, if a sample of individuals are classified according to two variables, say race and unemployment status, the resulting table of frequencies is called a *two-dimensional contingency table*, with each variable corresponding to one dimension of the table.

The analysis of relationships or associations between cross-classified variables has developed extensively over the last twenty years. For a comprehensive coverage of the analysis of cross-classified categorical data using traditional goodness-of-fit tests (Section 2.2) see, for example, Cox (1970), Haberman (1974, 1978, 1979), Bishop et al. (1975), Gokhale and Kullback (1978), Upton (1978), Fienberg (1980), Plackett (1981), Agresti (1984), Goodman (1984), and Freeman (1987).

3.2. Two-Dimensional Tables: Independence and Homogeneity

To illustrate the basic terminology and concepts for contingency tables, we consider the following example summarized from Kihlberg et al. (1964) (Fienberg, 1980, p. 90) comparing different types of car accidents. A total of 4,831 car accidents are classified according to variable 1: accident type (rollover, not rollover); and variable 2: accident severity (not severe, moderately severe, severe). Table 3.1 represents the resulting two-dimensional contingency table of frequencies.

In general, a two-dimensional table will have r rows representing the r categories of variable 1, denoted $A_1, A_2, \ldots, A_r$; and c columns for the c categories of variable 2, denoted $B_1, B_2, \ldots, B_c$. In the example described earlier, $r = 2$ with A_1 = rollover and A_2 = not rollover; $c = 3$ with B_1 = not severe, B_2 = moderately severe, and B_3 = severe. The resulting table will have rc cells and we define x_{ij} to be the number of observations classified as $A_i \cap B_j$, which is the cell frequency pertaining to the cell in row i and column j of the table. The *marginal* frequencies are the row totals and column totals obtained by summing the appropriate cell frequencies, and are denoted $x_{i+} = \sum_{j=1}^{c} x_{ij}$;

Table 3.1. Contingency Table for Car Accident Type by Accident Severity

	Accident severity			
Accident type	Not severe	Moderately severe	Severe	Total
Rollover	2,365	944	412	3,721
Not rollover	249	585	276	1,110
Total	2,614	1,529	688	4,831

Source: Kihlberg et al. (1964).

$i = 1, \ldots, r$ and $x_{+j} = \sum_{i=1}^{r} x_{ij}$; $j = 1, \ldots, c$. The total of the cell frequencies is denoted $n = x_{++} = \sum_{i=1}^{r}\sum_{j=1}^{c} x_{ij}$. Note that $n = \sum_{i=1}^{r} x_{i+} = \sum_{j=1}^{c} x_{+j}$. In general a "+" in place of a subscript denotes that summing has occurred over that subscript.

In order to study associations between the different classifying variables, we define π_{ij} to be the *probability* of observing an individual classified as $A_i \cap B_j$. Now we shall concentrate on building and testing different models for the π_{ij}.

Testing the Model of Independence

As an example, consider the accident data of Table 3.1. Is accident severity independent of whether the car rolls over or not? This model can be written as

$$Pr(\text{accident is } A_i \cap B_j) = Pr(\text{accident is } A_i) \cdot Pr(\text{accident is } B_j),$$

or, equivalently,

$$H_0\colon \pi_{ij} = \pi_{i+}\pi_{+j}, \tag{3.1}$$

where π_{i+} and π_{+j} represent the unknown marginal probabilities associated with an accident of type A_i and B_j, respectively, for $i = 1, \ldots, r$; $j = 1, \ldots, c$. Here the notation H_0 indicates that this is our hypothesized *null model* or *null hypothesis*, which we wish to test. Later we shall want to distinguish the null model from a hypothesized *alternative model* or *alternative hypothesis* (H_1), which is assumed to fit when the null model proves untenable.

Model (3.1) is called the model of *independence* or *no association*, and is the best-known model for contingency tables. In order to test the fit of this model to Table 3.1, we need to calculate the *expected* cell frequencies assuming the model. Here only the total sample size $n = 4{,}831$ is fixed. Let X_{ij} be the random variable associated with the observed cell frequency x_{ij}; $i = 1, \ldots, r$; $j = 1, \ldots, c$. Assume that the 4,831 accidents occur independently and each accident has probability π_{ij} of being in category $A_i \cap B_j$. Then it follows that $\mathbf{X} = (X_{11}, X_{12}, \ldots, X_{1c}, \ldots, X_{r1}, X_{r2}, \ldots, X_{rc})$ is distributed multinomially (defined by (2.4)) with sample size n and cell probabilities $\boldsymbol{\pi} = (\pi_{11}, \pi_{12}, \ldots, \pi_{1c}, \ldots, \pi_{r1}, \pi_{r2}, \ldots, \pi_{rc})$. Consequently the expected value of X_{ij} is $m_{ij} = n\pi_{ij}$, and we can rewrite hypothesis (3.1) in terms of the expected frequencies, as

$$H_0\colon m_{ij} = n\pi_{i+}\pi_{+j}. \tag{3.2}$$

In order to test the null hypothesis (3.2) for the data in Table 3.1, we still need to estimate each unknown marginal probability π_{i+}; $i = 1, \ldots, r$ and π_{+j}; $j = 1, \ldots, c$. In Section 3.4 we compare different methods of parameter estimation; but for now let us use the most obvious estimates, namely, the respective proportions x_{i+}/n and x_{+j}/n (these are also the *maximum likelihood estimates* or MLEs of π_{i+} and π_{+j}, respectively, as we shall see in Section 3.4). This gives us the *estimated* expected frequencies

$$\hat{m}_{ij} = n(x_{i+}/n)(x_{+j}/n) \\ = x_{i+}x_{+j}/n, \tag{3.3}$$

which are displayed for the accident data in Table 3.2.

Finally we propose testing the hypothesis (3.2) by using the *power-divergence family* of test statistics, defined from (2.17) as

$$2I^{\lambda}(\mathbf{x}:\hat{\mathbf{m}}) = \frac{2}{\lambda(\lambda+1)} \sum_{i=1}^{r} \sum_{j=1}^{c} x_{ij}\left[\left(\frac{x_{ij}}{\hat{m}_{ij}}\right)^{\lambda} - 1\right]; \qquad -\infty < \lambda < \infty, \tag{3.4}$$

where for $\lambda = -1, 0$, the limit as $\lambda \to -1, 0$ is used (Section 2.4). Table 3.3 illustrates some of the resulting calculations for different values of the family parameter λ. If the hypothesis (3.2) is true, and we consider $2I^{\lambda}(\mathbf{x}:\hat{\mathbf{m}})$ as a *random variable* with $\mathbf{x}$ replaced by the *random vector* $\mathbf{X}$ in both (3.3) and (3.4), then the (common) distribution of the power-divergence family members will be approximately chi-squared with degrees of freedom $(r-1)(c-1) = 2$. In Chapters 4 and 5, the distribution of the power-divergence statistic is discussed in detail under varying conditions (including those described earlier).

Now compare the values in Table 3.3 with those of the upper 95th percentile of a chi-squared distribution with two degrees of freedom, $\chi^2_2(0.05) = 5.99$. Regardless of which value of λ we use, there is very strong evidence that the statistic is not a realization from a chi-squared distribution with two degrees of freedom. Thus, we reject the null hypothesis H_0, and conclude that model (3.2) is inappropriate for these data. In other words, there appears to be a

Table 3.2. Estimated Expected Cell Frequencies Assuming the Model of Independence (3.2) for the Data in Table 3.1

	Accident severity			
Accident type	Not severe	Moderately severe	Severe	Total
Rollover	2,013.4	1,177.7	529.9	3,721
Not rollover	600.6	351.3	158.1	1,110
Total	2,614	1,529	688	4,831

Table 3.3. Computed Values of the Power-Divergence Statistic (3.4) under the Model of Independence (3.2) for the Data in Table 3.1 and Expected Frequencies in Table 3.2

λ									
-5	-2	-1	$-1/2$	0	1/2	2/3	1	2	5
2,092.2	784.1	662.8	626.3	602.1	588.1	585.5	583.2	598.0	857.1

strong association between accident severity and accident type. A closer comparison of Tables 3.1 and 3.2 indicates that rollover accidents result in less severe injuries than expected under the null model of independence (and accidents not involving rollover result in more severe injuries).

Testing the Model of Homogeneity

A variation of the model of independence discussed earlier occurs when the row totals are not random, but are fixed by the *sample design.* For example, in Table 3.4 (originally presented by Grizzle et al., 1969) four different operations for treating duodenal ulcer patients are compared in terms of the extent of one possible side effect, called *dumping syndrome*, described in Section 2.1. The operations are:

A_1 = drainage and vagotomy;
A_2 = 25% resection (antrechotomy) and vagotomy;
A_3 = 50% resection (hemigastrectomy) and vagotomy; and
A_4 = 75% resection.

In Section 2.1, the incidence of dumping syndrome after operation A_1 (drainage and vagotomy) was considered separately from the other three operations, all of which involve a partial gastrectomy (i.e., removal of part of the stomach). Now we have developed the necessary methodology to compare all four operations simultaneously.

In such a case we might consider the number of patients from each type of operation to be *fixed*; in other words, $x_{1+} = 96$ patients were sampled from operation A_1, and classified according to none, slight, or moderate dumping severity; and similarly for operations A_2, A_3, and A_4.

Unlike the car-accident example, where the sampling model was assumed to be the *full-multinomial* model, here we have four different multinomial

Table 3.4. Contingency Table for Dumping Severity by Operation

	Dumping severity			
Operation	None	Slight	Moderate	Total
A_1	61	28	7	96
A_2	68	23	13	104
A_3	58	40	12	110
A_4	53	38	16	107
Total	240	129	48	417

Reproduced from: J.E. Grizzle, C.F. Starmer and G.G. Koch, (1969). Analysis of categorical data by linear models. *Biometrics* **25**, 489–504. With permission from The Biometric Society.

samples of size x_{i+} ($i = 1, \ldots, 4$), with one from each operation. In this case, we are interested to see if the dumping severity is the same for all four operation types. That is,

$$Pr(\text{severity is } B_j | \text{operation is } A_i) = Pr(\text{severity is } B_j).$$

In other words, we are testing to see if the parameters of the $r = 4$ multinomial populations are the same. This model is more commonly referred to as the model of *homogeneity of proportions.*

In terms of expected cell frequencies, this model may be written

$$H_0\colon m_{ij} = x_{i+}\pi_{+j}, \tag{3.5}$$

where the common multinomial probability $\pi_{+j} = Pr(\text{severity is } B_j)$ is estimated (using *maximum likelihood*; see Section 3.4) by the marginal proportion x_{+j}/n for each severity type j. Therefore

$$\hat{m}_{ij} = x_{i+}x_{+j}/n, \tag{3.6}$$

which is identical to (3.3), the MLE used in the model of independence (3.2).

The sampling model just described is called the *product-multinomial* model in contrast to the *full-multinomial* model described for the car-accident data. Fortunately, as illustrated by (3.3) and (3.6), the MLEs for *independence* and *homogeneity* are algebraically equivalent. Therefore, it is not necessary to distinguish between the two sampling schemes for the purpose of computing the test statistics, unless a different estimation method is used for computing the expected cell frequencies for the model of homogeneity of proportions (Section 3.4). It turns out (Bishop et al., 1975, pp. 62–64) that the asymptotic distributions of the test statistics are also the same, viz., chi-squared with $(r-1)(c-1)$ degrees of freedom.

Table 3.5 gives the computed values of various power-divergence family members for the dumping-syndrome data. Assuming model (3.5) is true, the (common) distribution for the power-divergence family members will be approximately chi-squared with degrees of freedom $(r-1)(c-1) = 6$. Here the 90th and 95th percentiles are $\chi^2_6(0.10) = 10.65$ and $\chi^2_6(0.05) = 12.59$, respectively, so there is some question as to whether we should accept or reject the model of homogeneity given by (3.5). In Chapter 5, we shall see that the chi-squared distribution is an approximation of varying accuracy with λ. Furthermore, we shall see in Chapter 6 that the power-divergence statistic

Table 3.5. Computed Values of the Power-Divergence Statistic (3.4) under the Model of Homogeneity (3.5) for the Data in Table 3.4

λ									
−5	−2	−1	−1/2	0	1/2	2/3	1	2	5
15.31	11.98	11.35	11.09	10.87	10.69	10.64	10.54	10.32	10.23

measures different characteristics of the model fit depending on the value of λ. We delay resolving this example until Section 6.4. (As a postscript, it should be noted that partial gastrectomy surgery has now fallen somewhat into disuse, as the use of drugs has become a more popular treatment.)

Sampling Models

So far we have used sampling models based on the full-multinomial model (the car-accident data) where only the total sample size is fixed, and the product-multinomial model (the dumping-syndrome data) where both sample size and row margins are fixed. Another sampling model used with contingency tables is *Poisson* sampling. Here, neither the margins nor the total sample size n are fixed; we simply observe a set of independent Poisson processes, one for each cell in the table, over a fixed period of time. Further discussion of the full-multinomial, product-multinomial, and Poisson sampling models, together with some others, can be found in Bishop et al. (1975).

For the purpose of our discussion it is sufficient to recognize that, for all the loglinear models considered here, all three sampling models result in the same formulas for the estimated cell frequencies when using maximum likelihood estimation. Furthermore, for these hypothesized models, the goodness-of-fit test statistics are asymptotically equivalent under all three sampling models (Bishop et al. 1975, chapter 3).

3.3. Loglinear Models for Two and Three Dimensions

Both the model for independence (3.2) and the model for homogeneity (3.5) can be expressed in a linear form by taking logarithms of the expected cell frequencies, that is,

$$H_0\colon \log(m_{ij}) = \log(n) + \log(\pi_{i+}) + \log(\pi_{+j}), \tag{3.7}$$

and

$$H_0\colon \log(m_{ij}) = \log(n) + \log(x_{i+}/n) + \log(\pi_{+j}). \tag{3.8}$$

More generally, we can write

$$H_0\colon l_{ij} = \log(m_{ij}) = u + u_{1(i)} + u_{2(j)} \tag{3.9}$$

where

$$\begin{aligned} u &= l_{++}/rc, \\ u + u_{1(i)} &= l_{i+}/c, \\ u + u_{2(j)} &= l_{+j}/r. \end{aligned} \tag{3.10}$$

Note that $u_{1(+)} = u_{2(+)} = 0$ since these terms represent the deviations from u;

consequently there are $1 + (r - 1) + (c - 1) = r + c - 1$ parameters in model (3.9) and hence $rc - (r + c - 1) = (r - 1)(c - 1)$ degrees of freedom, just as in models (3.2) and (3.5). Furthermore, under product-multinomial sampling, the $u_{1(i)}$ term is fixed by the sample design for each i (see equation (3.8)) and is not otherwise interpretable.

Model (3.9) is called the *loglinear model* for independence (or homogeneity) for a two-dimensional contingency table (e.g., Bishop et al., 1975, p. 28). This form closely resembles the classical linear model notation for data assumed to be continuous and normally distributed; Haberman (1974) and McCullagh and Nelder (1983) illustrate some revealing parallels between these two classes of models.

The Saturated Model

As in the case of classical linear models, the most general loglinear model in two dimensions is

$$l_{ij} = u + u_{1(i)} + u_{2(j)} + u_{12(ij)}, \tag{3.11}$$

where, in addition to (3.10), we have added the *interaction* term

$$u_{12(ij)} = l_{ij} - l_{i+}/c - l_{+j}/r + l_{++}/rc.$$

Here $u_{12(ij)}$ indicates departure from independence (or homogeneity) and satisfies the constraints $u_{12(i+)} = u_{12(+j)} = 0$, for every i and j. The combination of these constraints, together with $u_{1(+)} = u_{2(+)} = 0$, indicates that (3.11) has $1 + (r - 1) + (c - 1) + (r - 1)(c - 1) = rc$ parameters and hence $rc - rc = 0$ degrees of freedom. A model such as this, in which the number of parameters equals the number of cells in the table, is called *saturated*.

Ordinal Models

Model (3.11) has $(r - 1)(c - 1)$ parameters more than model (3.9), and therefore represents a large jump in complexity from the very restrictive model of complete independence (or homogeneity). It makes sense to look for some middle ground between (3.9) and (3.11) wherein we try to specify a certain type of association between the variables. For example, if we can order the categories of both variables in some meaningful way, we may be able to detect a trend as we move from "low" to "high" categories. In the case of the dumping-syndrome data in Table 3.4, it is possible to order the severity of operations performed from A_1 being least severe to A_4 being most severe. In particular we can try fitting the ordinal loglinear model

$$H_0\colon l_{ij} = u + u_{1(i)} + u_{2(j)} + u'_{12}(v_i - \bar{v})(w_j - \bar{w}), \tag{3.12}$$

where v_i; $i = 1, \ldots, r$ and w_j; $j = 1, \ldots, c$ are scores ($\bar{v}$ and $\bar{w}$ are their averages) associated with the ordered categories of variables 1 and 2, respectively.

Agresti (1984) provides a very thorough discussion of such ordinal models and shows how well (3.12) fits the dumping-syndrome data even though this model has only one more parameter, namely u'_{12}, than model (3.9). See also Goodman (1984) for a detailed discussion of models for data with ordered categories.

Three Dimensions

Following the analogy with classical linear models, the generalization to the case of cross-classifying three variables into a three-dimensional contingency table is straightforward. The most general (i.e., saturated) model is

$$l_{ijl} = \log(m_{ijl}) = u + u_{1(i)} + u_{2(j)} + u_{3(l)} + u_{12(ij)} + u_{13(il)} + u_{23(jl)} + u_{123(ijl)}, \tag{3.13}$$

where we assume m_{ijl} is the expected cell frequency in the ith row, jth column and lth layer of the table. As in the two-dimensional case there are several constraints on the terms, namely,

$$u_{1(+)} = u_{2(+)} = u_{3(+)} = 0,$$

$$u_{12(i+)} = u_{12(+j)} = u_{13(i+)} = u_{13(+l)} = u_{23(j+)} = u_{23(+l)} = 0,$$

and

$$u_{123(ij+)} = u_{123(i+l)} = u_{123(+jl)} = 0.$$

The simplest model, by analogy with the independence model in two dimensions, is

$$H_0\colon l_{ijl} = u + u_{1(i)} + u_{2(j)} + u_{3(l)}, \tag{3.14}$$

where we assume complete independence of variables 1, 2, and 3. When this model does not fit the data, there are three other classes of models we can try. These are:

(a) Variables 1 and 2 are dependent on each other, but are jointly independent of variable 3:

$$H_0\colon l_{ijl} = u + u_{1(i)} + u_{2(j)} + u_{3(l)} + u_{12(ij)}. \tag{3.15}$$

(b) Variables 1 and 2 are dependent and variables 1 and 3 are dependent, but variables 2 and 3 are conditionally independent given variable 1:

$$H_0\colon l_{ijl} = u + u_{1(i)} + u_{2(j)} + u_{3(l)} + u_{12(ij)} + u_{13(il)}. \tag{3.16}$$

Note that interchanging variables 1, 2, and 3 leads to two more versions of both (3.15) and (3.16).

(c) Variables 1, 2, and 3 interact pairwise, but there is no three-factor interaction:

$$H_0\colon l_{ijl} = u + u_{1(i)} + u_{2(j)} + u_{3(l)} + u_{12(ij)} + u_{13(il)} + u_{23(jl)}. \tag{3.17}$$

Estimation of the parameters for models (3.14) through (3.17) is discussed in Section 3.4, and strategies for model selection are described briefly in Section 3.6.

Other Approaches to Modeling

Various other approaches have been considered for analyzing cross-classified data, including models linear in the cell probabilities (Section 3.5; Bishop et al., 1975, p. 23) and *correspondence analysis* (e.g., Benzécri, 1973; Goodman, 1986). However the loglinear model is by far the most popular. Its advantages over other parameterizations include: (a) a direct link with the concept of independence of cell probabilities; (b) accessibility of the many algebraic identities available for classical linear models; and (c) widely available computer packages for data analysis and model selection.

Other models that incorporate the ordinal nature of any of the three variables can be defined analogously to (3.12); see Agresti (1984) and Goodman (1984).

3.4. Parameter Estimation Methods: Minimum Distance Estimation

Once we have hypothesized a model for the data, we need to estimate any unknown parameters before we can calculate the expected cell frequencies.

Maximum Likelihood Estimation

The best-known method of parameter estimation is the *method of maximum likelihood*. In particular, using the notation for a two-dimensional contingency table, the loglikelihood under full-multinomial, product-multinomial, and Poisson sampling schemes can be shown to be proportional to

$$\sum_{i=1}^{r} \sum_{j=1}^{c} x_{ij} \log(m_{ij}/x_{ij}), \tag{3.18}$$

for the models of interest. Maximum likelihood estimation requires a constrained maximization of (3.18) with respect to each expected frequency m_{ij}, using the constrants imposed by the hypothesized model and sample design (clearly if m_{ij} is constrained only by $m_{++} = x_{++} = n$, then $\hat{m}_{ij} = x_{ij}$, for each i and j, will maximize (3.18)). For the model of independence or homogeneity given by (3.9), it is straightforward to show that the MLE is $\hat{m}_{ij} = x_{i+}x_{+j}/n$, which is used in (3.3) and (3.6).

Similarly, for the three-dimensional models (3.14) through (3.17), it is possible to derive the following MLEs for the expected cell frequencies (Bishop et al., 1975, chapter 3): for the model (3.14)

$$\hat{m}_{ijl} = x_{i++}x_{+j+}x_{++l}/n^2, \tag{3.19}$$

for the model (3.15)

$$\hat{m}_{ijl} = x_{ij+}x_{++l}/n, \tag{3.20}$$

and for the model (3.16)

$$\hat{m}_{ijl} = x_{ij+}x_{i+l}/x_{i++}. \tag{3.21}$$

For the model (3.17), we cannot write a closed-form expression for the estimated expected frequencies. One way to compute these MLEs is to use an iterative procedure whereby the m_{ijl} are progressively constrained by the requirements of (3.17). One such procedure is called *iterative proportional fitting* and is the algorithm used by many computer packages for calculating expected frequencies for loglinear models. A simple descriptive example of iterative proportional fitting is given by Fienberg (1980, chapter 3); see also Bishop et al., (1975, section 3.5). An alternative approach based on the Newton-Raphson method is described by Haberman (1974, 1978, 1979). Denteneer and Verbeek (1986) provide a series of comprehensive efficiency comparisons for various implementations of these algorithms.

Minimum Power-Divergence Estimation

For the three sampling schemes we are considering here, it is clear that the method of maximum likelihood (based on (3.18)) is equivalent to minimizing the loglikelihood ratio statistic

$$G^2 = 2\sum_{i=1}^{r}\sum_{j=1}^{c} x_{ij}\log(x_{ij}/m_{ij}), \tag{3.22}$$

which we have already demonstrated in Section 2.4 to be a member of the power-divergence family of statistics. There, the sampling model considered was the full-multinomial model. In order to take into account the product-multinomial and Poisson sampling models, and to allow for multidimensional contingency tables, we now rewrite the power-divergence family of (3.4) as

$$2I^{\lambda}(\mathbf{x} : \hat{\mathbf{m}}) = \frac{2}{\lambda(\lambda+1)}\sum_{i=1}^{k} x_i\left[\left(\frac{x_i}{\hat{m}_i}\right)^{\lambda} - 1\right]; \qquad -\infty < \lambda < \infty, \tag{3.23}$$

where for $\lambda = -1, 0$, the limit as $\lambda \to -1, 0$ is used. The subscript i runs over all possible cells in the contingency table (by stringing all the columns end to end) and k represents the total number of cells in the table.

Now define the general null model for $\mathbf{m}$ to be

$$H_0\colon \mathbf{m} \in M_0, \tag{3.24}$$

where $M_0 \subset M = \{$all possible cell frequency vectors consistent with the margins fixed by the sample design$\}$ defines the subset of all frequency vectors satisfying the constraints of the hypothesized model. When presented in this way, the method of maximum likelihood is seen to be equivalent to minimizing (3.23) with respect to $\mathbf{m} \in M_0$ using $\lambda = 0$. A natural generalization is to consider minimizing (3.23) using different (fixed) values of λ. This leads us to define the *minimum power-divergence estimate* of $\mathbf{m} \in M_0$ to be the $\hat{\mathbf{m}}^{(\lambda)}$ satisfying

$$I^{\lambda}(\mathbf{x} : \hat{\mathbf{m}}^{(\lambda)}) = \inf_{\mathbf{m} \in M_0} I^{\lambda}(\mathbf{x} : \mathbf{m}); \qquad -\infty < \lambda < \infty, \tag{3.25}$$

where $I^{\lambda}(\mathbf{x} : \mathbf{m})$ is defined from (3.23) by replacing $\hat{\mathbf{m}}$ with $\mathbf{m}$. The strict convexity of $I^{\lambda}(\mathbf{x} : \mathbf{m})$ ensures uniqueness of the estimate $\hat{\mathbf{m}}^{(\lambda)}$.

Other estimates (less well known than the MLE) which are members of the family of minimum power-divergence estimates include: the minimum chi-squared estimate (Neyman, 1949) for $\lambda = 1$; the minimum modified chi-squared estimate (Neyman, 1949) for $\lambda = -2$; the modified MLE or *minimum discrimination information* estimate for the *external constraints problem* (discussed in Section 3.5; see also Kullback, 1985) for $\lambda = -1$; and the minimum Matusita distance (or Hellinger distance) estimate (Matusita, 1954) for $\lambda = -1/2$. All of these estimates fall into a class called *minimum distance estimates* (even though $I^{\lambda}(\mathbf{x} : \mathbf{m})$ is not a true distance function between $\mathbf{x}$ and $\mathbf{m}$, but is rather a directed divergence as discussed in Section 7.4). Parr (1981) provides an extensive bibliography for minimum distance estimates.

Weighted Least Squares and the Power-Divergence Estimate

Consider a null model for the cell probabilities $\boldsymbol{\pi} = (\pi_1, \pi_2, \ldots, \pi_k)$ given by the equations

$$g_j(\boldsymbol{\pi}) = \boldsymbol{\beta}\mathbf{w}_j'; \qquad j = 1, \ldots, t,$$

where each $g_j(\boldsymbol{\pi})$ is a smooth function on the probability simplex, $\mathbf{w}_j = (w_{j1}, w_{j2}, \ldots, w_{ju})$ is a vector of constants and $\boldsymbol{\beta} = (\beta_1, \beta_2, \ldots, \beta_u)$ is a vector of (unknown) parameters. In matrix notation, this becomes

$$\mathbf{g}(\boldsymbol{\pi}) = \boldsymbol{\beta}W', \tag{3.26}$$

where $\mathbf{g}(\boldsymbol{\pi}) = (g_1(\boldsymbol{\pi}), g_2(\boldsymbol{\pi}), \ldots, g_t(\boldsymbol{\pi}))$ and W is the $t \times u$ design matrix with jth row given by $\mathbf{w}_j$. Hence we assume that after transformation by $\mathbf{g}$, the null model is linear. Grizzle et al. (1969) propose estimating $\boldsymbol{\beta}$ in (3.26) by the method of *weighted least squares*. In other words, by minimizing

$$(\mathbf{g}(\mathbf{x}/n) - \boldsymbol{\beta}W')S^{-}(\mathbf{g}(\mathbf{x}/n) - \boldsymbol{\beta}W')' \tag{3.27}$$

with respect to $\boldsymbol{\beta}$, where S^{-} is a generalized inverse of the sample covariance matrix of $\mathbf{g}(\mathbf{x}/n)$. When $\mathbf{g}$ is a linear function, S can be written down exactly;

when $\mathbf{g}$ is nonlinear, S is obtained approximately by the δ-method. A comprehensive overview of the weighted least squares approach is given by Forthofer and Lehnen (1981); further examples and comparisons with maximum likelihood can be found in Freeman (1987).

When the functions $g_j(\boldsymbol{\pi})$ are linear in the cell probabilities, it is well known that the weighted least squares approach is algebraically equivalent to the minimum modified chi-squared approach due to Neyman (1949) (i.e., the minimum power-divergence estimation approach with $\lambda = -2$; see (3.25)). When the functions $g_j(\boldsymbol{\pi})$ are nonlinear, the two approaches are equivalent only after linearization using a Taylor-series expansion (e.g., Bishop et al., 1975, section 10.3).

We now derive a general relationship between the minimum power-divergence approach defined by (3.25) and the weighted least squares approach. In the simplest case, consider the power transformation

$$\mathbf{g}^{\omega}(\boldsymbol{\pi}) \equiv \left[\frac{\pi_1^{\omega} - 1}{\omega}, \frac{\pi_2^{\omega} - 1}{\omega}, \ldots, \frac{\pi_k^{\omega} - 1}{\omega}\right].$$

The approximate covariance matrix of $\mathbf{g}^{\omega}(\mathbf{x}/n)$ is derived easily from the δ-method. Using a very convenient generalized inverse, (3.27) becomes

$$n \sum_{i=1}^{k} \left[\frac{(x_i/n)^{\omega} - 1}{\omega} - \frac{\pi_i^{\omega} - 1}{\omega}\right]^2 (x_i/n)^{1-2\omega}, \tag{3.28}$$

where

$$[(\pi_1^{\omega} - 1)/\omega, (\pi_2^{\omega} - 1)/\omega, \ldots, (\pi_k^{\omega} - 1)/\omega] = \boldsymbol{\beta} W'. \tag{3.29}$$

In Section 6.6 we shall see that (3.28) can be approximated by $2nI^{\omega}(\boldsymbol{\pi} : \mathbf{x}/n) = 2nI^{-\omega-1}(\mathbf{x}/n : \boldsymbol{\pi})$. Therefore the weighted least squares estimate obtained by minimizing (3.28) will be close to the minimum power-divergence estimate $\hat{\boldsymbol{\pi}}^{(\lambda)}$ with $\lambda = -\omega - 1$ (obtained by minimizing $2nI^{\lambda}(\mathbf{x}/n : \boldsymbol{\pi})$ subject to (3.29)).

For the special case $\omega = 1$ the model (3.29) is linear in the cell probabilities. The respective minimum power-divergence estimate is $\hat{\boldsymbol{\pi}}^{(-\omega-1)} = \hat{\boldsymbol{\pi}}^{(-2)}$ which is Neyman's minimum modified chi-squared estimate, as it should be.

The literature and commercial computer packages for the weighted least squares approach usually limit the $\mathbf{g}$ functions (3.26) to the linear, log, logistic, or exponential (e.g., Forthofer and Lehnen, 1981; the procedure CATMOD in SAS, 1982). We would recommend that these should be extended to include the power transformation (see Section 8.2 for further discussion).

Comparing Estimators

If we replace the observed frequencies $\mathbf{x}$ in (3.25) by the random vector $\mathbf{X}$, then $\hat{\mathbf{m}}^{(\lambda)}$ becomes a random vector. We call this random vector an *estimator*, which yields the appropriate estimate when $\mathbf{X}$ is replaced by any specific observed vector $\mathbf{x}$. Maximum likelihood estimation is often justified through the large-sample properties of the *maximum likelihood estimator* such as consistency and efficiency. Neyman (1949) defines a class of *best asymptotically normal* (BAN) estimators, all of which share the same optimal large-sample properties. In Section A5 of the Appendix we prove that the minimum power-divergence estimator is in this class, provided the regularity conditions of Birch (1964) hold.

Why should we consider alternatives to the MLE? At first sight there appears to be little reason to deviate from the traditional maximum likelihood path, and in fact Rao (1961, 1962) defines a *second-order efficiency* criterion for which he shows the method of maximum likelihood to provide the unique optimum estimate. However, Rao's second-order efficiency criterion has been called into question by a number of authors (e.g., Berkson, 1980; Parr, 1981; Causey, 1983; Harris and Kanji, 1983), and the alternative criterion of Hodges-Lehmann deficiency yields quite different results (Section A1 of the Appendix). In general, it seems quite reasonable to recommend matching the minimum power-divergence estimate with the choice of λ in the power-divergence test statistic. For example, if it is decided to use $\lambda = 2/3$ for the test statistic (3.23), then we might estimate $\mathbf{m}$ by $\hat{\mathbf{m}}^{(2/3)}$. Therefore the MLE $\hat{\mathbf{m}}^{(0)}$ would be recommended when using the loglikelihood ratio statistic $G^2(\lambda = 0)$. From the computational point of view, all that is needed is a general minimization routine to compute the estimate (e.g., Böhning and Holling, 1988). However if only the MLE $\hat{\mathbf{m}}^{(0)}$ is readily available, we would have no hesitation in using it in the computation of, say, the test statistic $2I^{2/3}(\mathbf{x} : \hat{\mathbf{m}}^{(0)})$, since it is asymptotically equivalent to $2I^{2/3}(\mathbf{x} : \hat{\mathbf{m}}^{(2/3)})$ ($\hat{\mathbf{m}}^{(0)}$ is BAN).

An Example of Minimum Power-Divergence Estimation

As an example, we return to the dumping-syndrome data in Table 3.4, and the test for homogeneity of proportions. Following the results of Quade and Salama (1975), it is straightforward to deduce that the minimum power-divergence estimate from (3.25) is

$$\hat{m}_{ij}^{(\lambda)} = \frac{x_{i+}\left[\sum_{i=1}^{r} x_{ij}^{\lambda+1}/x_{i+}^{\lambda}\right]^{1/(\lambda+1)}}{\sum_{j=1}^{c}\left[\sum_{i=1}^{r} x_{ij}^{\lambda+1}/x_{i+}^{\lambda}\right]^{1/(\lambda+1)}}. \tag{3.30}$$

For $\lambda = 1$, (3.30) becomes

$$\hat{m}_{ij}^{(1)} = \frac{x_{i+}\left[\sum_{i=1}^{r} x_{ij}^{2}/x_{i+}\right]^{1/2}}{\sum_{j=1}^{c}\left[\sum_{i=1}^{r} x_{ij}^{2}/x_{i+}\right]^{1/2}}, \tag{3.31}$$

which is the minimum chi-squared estimate. For $\lambda = 0$, (3.30) reduces to

$$\hat{m}_{ij}^{(0)} = x_{i+}x_{+j}/n, \tag{3.32}$$

the standard MLE as in (3.6). For $\lambda = -2$, (3.30) becomes

$$\hat{m}_{ij}^{(-2)} = \frac{x_{i+}\left[\sum_{i=1}^{r} x_{i+}^2/x_{ij}\right]^{-1}}{\sum_{j=1}^{c}\left[\sum_{i=1}^{r} x_{i+}^2/x_{ij}\right]^{-1}}, \tag{3.33}$$

which is the minimum modified chi-squared estimate. Table 3.6 gives the estimated expected frequencies for Table 3.4, using (3.31), (3.32), and (3.33).

From Table 3.6 we see that the marginal totals for the expected frequencies using $\lambda = 1$ and $\lambda = -2$, no longer equal the observed marginal totals as in the case $\lambda = 0$. In this example, the different estimates of each expected cell frequency are all very similar. Table 3.7 gives the values of $2I^\lambda(\mathbf{x} : \hat{\mathbf{m}}^{(\lambda)})$ for various λ values, where the same value of λ is used for estimating the cell frequencies and testing the hypothesis. From the definition of $\hat{\mathbf{m}}^{(\lambda)}$ in (3.25), $2I^\lambda(\mathbf{x} : \mathbf{m})$ is minimized under the null model when the expected cell frequencies are estimated by $\hat{\mathbf{m}}^{(\lambda)}$, and so all the values in Table 3.7 are less than or equal to the corresponding values in Table 3.5.

Table 3.6. Estimated Expected Cell Frequencies for Data in Table 3.4 and the Model of Homogeneity of Proportions, Using the Minimum Power-Divergence Estimate with $\lambda = 1$, 0, and -2

	Dumping Severity			
Operation	None	Slight	Moderate	Total
A_1	54.95*	29.84	11.22	96.0
	55.25	29.70	11.05	96.0
	56.03	29.34	10.63	96.0
A_2	59.53	32.32	12.15	104.0
	59.86	32.17	11.97	104.0
	60.70	31.78	11.52	104.0
A_3	62.96	34.19	12.85	110.0
	63.31	34.03	12.66	110.0
	64.20	33.62	12.18	110.0
A_4	61.24	33.25	12.50	107.0
	61.58	33.10	12.32	107.0
	62.45	32.70	11.85	107.0
Total	238.7	129.6	48.7	417.0
	240.0	129.0	48.0	417.0
	243.4	127.4	46.2	417.0

* Each triple refers to expected frequencies using $\lambda = 1$ (3.31), $\lambda = 0$ (3.32), and $\lambda = -2$ (3.33), respectively.

Table 3.7. Recomputed Values of the Power-Divergence Statistic in Table 3.5 Using the Respective Minimum Power-Divergence Estimates Instead of MLEs

				λ					
-5	-2	-1	$-1/2$	0	1/2	2/3	1	2	5
13.88	11.85	11.32	11.09	10.87	10.69	10.63	10.52	10.25	9.9

Note that in both Table 3.5 and Table 3.7, the value of the power-divergence statistic decreases as λ increases. In Chapter 6 we shall see that the various λ emphasize different types of deviation from the model. Consequently, our choice of λ will depend on the type of relative deviations between the observed and expected cell frequencies we wish to detect. We defer further discussion of this important point to Sections 6.1 and 6.2.

3.5. Model Generation: A Characterization of the Loglinear, Linear, and Other Models through Minimum Distance Estimation

In Section 3.4 we noted that estimation based on *minimum discrimination information* (MDI) (Kullback, 1959; Gokhale and Kullback, 1978; Kullback and Keegel, 1984) can be used to estimate expected cell frequencies in contingency tables, and is a special case (viz., $\lambda = -1$) of the more general minimum power-divergence estimate. However the principle of MDI estimation can be used further to derive the general loglinear model for the cell frequencies. In this section we give an overview of the principle of MDI estimation and provide a simple generalization using the power-divergence statistic, which allows us to generate characterizations of models other than the loglinear, including the *linear* or *additive* model.

We define the discrimination information for the probability vectors $\boldsymbol{\pi} = (\pi_1, \pi_2, \ldots, \pi_k)$ and $\mathbf{p} = (p_1, p_2, \ldots, p_k)$ (with $\pi_+ = p_+ = 1$) as

$$K(\boldsymbol{\pi} : \mathbf{p}) = \sum_{i=1}^{k} \pi_i \log(\pi_i/p_i) \tag{3.34}$$

(Kullback, 1959). The discrimination information measures the divergence between $\boldsymbol{\pi}$ and $\mathbf{p}$ with $K(\boldsymbol{\pi} : \mathbf{p}) = 0$ if and only if $\pi_i = p_i$ for all $i = 1, \ldots, k$ (Kullback, 1983, provides a description of its properties, some of which are described in Section 7.4).

The concept behind MDI estimation is that we have a set of linear constraints that a probability vector $\boldsymbol{\pi}$ must satisfy. Among all $\boldsymbol{\pi}$ that satisfy these constraints, we wish to find the one that most closely resembles some fixed a priori vector $\mathbf{p}$, in the sense that it minimizes $K(\boldsymbol{\pi} : \mathbf{p})$. As an example, for contingency tables these constraints might be on the marginal frequencies, and $\mathbf{p}$ might be the simplest possible vector of equal probabilities (i.e., $\mathbf{p} = \mathbf{1}/k = (1/k, 1/k, \ldots, 1/k)$) as we shall see later.

In other words, given a set of $s + 1$ linear constraints on $\boldsymbol{\pi}$,

$$\sum_{i=1}^{k} c_{ij}\pi_i = \theta_j; \qquad j = 0, 1, \ldots, s \tag{3.35}$$

and an a priori vector $\mathbf{p}$, we define the MDI estimate $\boldsymbol{\pi}^*$ as the value of $\boldsymbol{\pi}$ which minimizes $K(\boldsymbol{\pi} : \mathbf{p})$ subject to satisfying (3.35). Throughout, we define $c_{i0} = 1$ for $i = 1, \ldots, k$ and $\theta_0 = 1$ so as to include the natural constraint $\pi_+ = 1$ in (3.35).

Solving this minimization implies that $\boldsymbol{\pi}^*$ has the loglinear form

$$\log(\pi_i^*/p_i) = \tau_0 + c_{i1}\tau_1 + c_{i2}\tau_2 + \cdots + c_{is}\tau_s, \tag{3.36}$$

for $i = 1, \ldots, k$ (e.g., Gokhale and Kullback, 1978; Section A2 of the Appendix contains a proof of the more general result given by (3.45)). The strict convexity of $K(\boldsymbol{\pi} : \mathbf{p})$ ensures that $\boldsymbol{\pi}^*$ is unique, and the existence of $\boldsymbol{\pi}^*$ is assured as a consequence of the *iterative proportional fitting* algorithm (e.g., Darroch and Speed, 1983).

An important partitioning of $K(\boldsymbol{\pi} : \mathbf{p})$ to which we shall refer later is

$$K(\boldsymbol{\pi} : \mathbf{p}) = K(\boldsymbol{\pi} : \boldsymbol{\pi}^*) + K(\boldsymbol{\pi}^* : \mathbf{p}) \tag{3.37}$$

which holds for every $\boldsymbol{\pi}$ satisfying (3.35). This equation is the basis for partitioning goodness-of-fit statistics into analysis-of-information tables, which parallel the analysis-of-variance tables in classical linear model theory (Gokhale and Kullback, 1978). (In Section 7.4 another important parallel to analysis of variance is described, which is called analysis of diversity; see Rao, 1982b; 1986.)

Internal Constraints Problems (ICP)

There are two broad classes of problems in categorical data analysis to which the principle of MDI estimation can be specialized (Kullback, 1983; Kullback and Keegel, 1984). In the first class, the objective is to smooth the data or build a model based on the data. Here the constraints (3.35) are derived internally from the observed data; consequently the class of problems is called *internal constraints problems* (ICP). If we define the expected frequencies by $\mathbf{m} = n\boldsymbol{\pi}$, and let $\mathbf{x}$ represent the observed contingency table, then the constraints (3.35) for ICP are written

$$\sum_{i=1}^{k} c_{ij}m_i = \sum_{i=1}^{k} c_{ij}x_i; \qquad j = 0, 1, \ldots, s. \tag{3.38}$$

These constrain certain expected margins to equal the respective observed margins (in particular $c_{i0} = 1$; $i = 1, \ldots, k$, which gives $m_+ = x_+$). For ICP, the a priori probability vector $\mathbf{p}$ is usually taken to be the equiprobable vector $\mathbf{p} = \mathbf{1}/k$. Consequently the MDI estimate $\mathbf{m}^* = n\boldsymbol{\pi}^*$ is the "simplest" vector $\mathbf{m}$ satisfying (3.38) in the sense that $\boldsymbol{\pi}^*$ minimizes $K(\boldsymbol{\pi} : \mathbf{1}/k)$ among all vectors $\boldsymbol{\pi} = \mathbf{m}/n$ with $\mathbf{m}$ satisfying (3.38). At the end of this section we give an example where $\mathbf{p} \neq \mathbf{1}/k$.

For ICP it is straightforward to show that the MDI estimate $\mathbf{m}^*$ is also the MLE for the loglinear model with constraints (3.38).

Gokhale and Kullback (1978) and Kullback (1985) recommend that the appropriate goodness-of-fit statistic for ICP is

$$2K(\mathbf{x} : \mathbf{m}^*) = 2 \sum_{i=1}^{k} x_i \log(x_i/m_i^*). \tag{3.39}$$

Note that now the definition of $K(\mathbf{x} : \mathbf{m}^*)$ refers to frequency vectors rather than the probability vectors of (3.34). There should be no confusion with this slight abuse of notation. We have already done the same with the more general power-divergence statistic (equations (2.16) and (2.17)). When the observed vector $\mathbf{x}$ is replaced by the multinomial random vector $\mathbf{X}$ in both (3.38) and (3.39), then the statistic (3.39) has an asymptotic chi-squared distribution with $k - s - 1$ degrees of freedom, provided the hypothesized set of constraints (3.38) correctly describes the distributional parameters of $\mathbf{X}$ (Gokhale and Kullback, 1978; see also Section 4.1 for a more general result in the case of the power-divergence statistic). This is the loglikelihood ratio statistic G^2 from (2.10) (i.e., the power-divergence statistic with $\lambda = 0$).

Gokhale and Kullback's choice of (3.39) as a test statistic stems from their interest in partitioning the information from nested models. Specifically, if $\mathbf{m}_a^*$ is the MDI estimated frequency vector from one set of $s_a + 1$ (internal) constraints and $\mathbf{m}_b^*$ is the estimated frequency vector from another set of $s_b + 1$ (internal) constraints which include those for $\mathbf{m}_a^*$ (i.e., we have nested models) then we obtain the partition

$$2K(\mathbf{x} : \mathbf{m}_a^*) = 2K(\mathbf{m}_b^* : \mathbf{m}_a^*) + 2K(\mathbf{x} : \mathbf{m}_b^*). \tag{3.40}$$

This follows by analogy with (3.37) (if we replace the probability vectors with frequency vectors) by observing that since $\mathbf{x}$ will always satisfy internal constraints of the type (3.38),

$$K(\mathbf{x} : n\mathbf{1}/k) = K(\mathbf{x} : \mathbf{m}_a^*) + K(\mathbf{m}_a^* : n\mathbf{1}/k),$$

$$K(\mathbf{x} : n\mathbf{1}/k) = K(\mathbf{x} : \mathbf{m}_b^*) + K(\mathbf{m}_b^* : n\mathbf{1}/k),$$

and finally, since $\mathbf{m}_b^*$ satisfies the constraints for $\mathbf{m}_a^*$, then

$$K(\mathbf{m}_b^* : n\mathbf{1}/k) = K(\mathbf{m}_b^* : \mathbf{m}_a^*) + K(\mathbf{m}_a^* : n\mathbf{1}/k).$$

Subtracting the first two equations and using the expression for $K(\mathbf{m}_b^* : \mathbf{m}_a^*)$ from the third gives (3.40). Provided the less restrictive set of hypothesized constraints associated with $\mathbf{m}_a^*$ is correct (and $\mathbf{x}$ is replaced by $\mathbf{X}$), then all three statistics in (3.40) have asymptotic chi-squared distributions with degrees of freedom $k - s_a - 1$, $s_b - s_a$, and $k - s_b - 1$, respectively.

External Constraints Problems (ECP)

In the second broad class of problems to which the principle of MDI estimation can be applied, we consider situations in which the objective is not to smooth the data but rather to test some external hypothesis (which is not derived from the data). This class of problems is called *external constraints problems* (ECP). Examples of ECP for contingency tables include: tests for symmetry, tests for marginal homogeneity, tests of specified marginal frequencies, and tests of linear or additive (rather than loglinear) models (e.g., Gokhale and Kullback, 1978). For ECP, the MDI estimation approach and the maximum likelihood approach differ.

The ECP constraints (3.35) take the form

$$\sum_{i=1}^{k} c_{ij} m_i = n\theta_j; \qquad j = 0, 1, \ldots, s \tag{3.41}$$

where θ_j are fixed (with $c_{i0} = 1$; $i = 1, \ldots, k$ and $\theta_0 = 1$ so that $m_+ = n$). The objective is to find the value $\mathbf{m}^*$ among those $\mathbf{m}$ satisfying the constraints (3.41) that most closely resembles the observed contingency table $\mathbf{x}$. Consequently for ECP, the a priori vector $\mathbf{p}$ in (3.34) is usually taken to be the observed proportions $\mathbf{x}/n$, implying that $\mathbf{m}^*/n = \boldsymbol{\pi}^*$ is closest to the observed proportions $\mathbf{x}/n$ (in terms of $K(\boldsymbol{\pi} : \mathbf{x}/n)$ where $\boldsymbol{\pi} = \mathbf{m}/n$) among all vectors $\mathbf{m}$ satisfying (3.41).

For ECP, Gokhale and Kullback (1978) and Kullback (1985) recommend that the appropriate goodness-of-fit statistic is

$$2K(\mathbf{m}^* : \mathbf{x}) = 2 \sum_{i=1}^{k} m_i^* \log(m_i^*/x_i). \tag{3.42}$$

When the observed vector $\mathbf{x}$ is replaced by the multinomial random vector $\mathbf{X}$ in (3.42) (and in calculating $\mathbf{m}^*$), then this statistic has an asymptotic chi-squared distribution with s degrees of freedom, provided the hypothesized set of constraints (3.41) correctly describe the parameters of the distribution of $\mathbf{X}$. This is the minimum discrimination information statistic GM^2 from (2.12) (i.e., the power-divergence statistic with $\lambda = -1$). Once again the choice of the test statistic is based on the partitioning property

$$2K(\mathbf{m}_b^* : \mathbf{x}) = 2K(\mathbf{m}_b^* : \mathbf{m}_a^*) + 2K(\mathbf{m}_a^* : \mathbf{x}) \tag{3.43}$$

which follows by direct analogy with (3.37), where $\mathbf{m}_a^*$ and $\mathbf{m}_b^*$ are defined as for (3.40) (with, respectively, $s_a + 1$ and $s_b + 1$ external constraints of the type (3.41)). Provided the more restrictive set of hypothesized constraints associated with $\mathbf{m}_b^*$ is correct (and $\mathbf{x}$ is replaced by $\mathbf{X}$), then all three statistics in

(3.43) have asymptotic chi-squared distributions with degrees of freedom s_b, $s_b - s_a$, and s_a, respectively.

Examples of the application of minimum discrimination information (MDI) estimation to both ICP (3.38) and ECP (3.41) can be found in Gokhale and Kullback (1978) and Kullback and Keegel (1984).

MDI Estimation—A Special Case of Minimum Power-Divergence Estimation

Minimum discrimination information (MDI) estimation for both ICP and ECP are special cases of the minimum power-divergence estimation procedure described in Section 3.4. For the ICP we have already noted that MDI estimation is equivalent to maximum likelihood estimation or minimum power-divergence estimation defined in (3.25) with $\lambda = 0$; the goodness-of-fit statistic is simply $2I^0(\mathbf{x} : \hat{\mathbf{m}})$ where $\hat{\mathbf{m}}$ is the MLE (or minimum power-divergence estimate $\hat{\mathbf{m}}^{(0)}$) of the expected frequencies based on the "smoothing" constraints (3.38). For the ECP the goodness-of-fit statistic $2K(\mathbf{m}^* : \mathbf{x})$ is equivalent to $2I^{-1}(\mathbf{x} : \mathbf{m}^*)$, where $\mathbf{m}^*$ is just the minimum power-divergence estimate $\hat{\mathbf{m}}^{(-1)}$ and the hypothesized null model H_0 is defined by (3.41).

Generalizing the Principle of MDI Estimation

Following the principle of MDI estimation for ICP, we now define a more general principle we call *generalized minimum power-divergence estimation* by paralleling some results from Darroch (1976) and Darroch and Speed (1983).

Assume we have a fixed probability vector $\mathbf{p}$, a value of λ, and a set of $s + 1$ constraints

$$\sum_{i=1}^{k} c_{ij} m_i = n\theta_j; \qquad j = 0, 1, \ldots, s, \tag{3.44}$$

with $c_{i0} = 1$ for $i = 1, \ldots, k$ and $\theta_0 = 1$ (i.e., $m_+ = n$). If there exists a vector $\mathbf{m}^{*(\lambda)}$ satisfying (3.44), which can be written in the form

$$\frac{1}{\lambda}\left[\left(\frac{m_i^{*(\lambda)}}{np_i}\right)^{\lambda} - 1\right] = \tau_0 + c_{i1}\tau_1 + c_{i2}\tau_2 + \cdots + c_{is}\tau_s, \tag{3.45}$$

then we call $\mathbf{m}^{*(\lambda)}$ the *generalized minimum power-divergence estimate*, since (as is shown in Section A2 of the Appendix) it is the vector closest to $n\mathbf{p}$ in the sense of minimizing

$$2I^{\lambda}(\mathbf{m} : n\mathbf{p}) = \frac{2}{\lambda(\lambda+1)} \sum_{i=1}^{k} m_i\left[\left(\frac{m_i}{np_i}\right)^{\lambda} - 1\right]; \qquad -\infty < \lambda < \infty. \tag{3.46}$$

The power-divergence measure (3.46) is analogous to (2.17). We now consider some special cases.

The principle of MDI estimation is obtained by setting $\lambda = 0$ in (3.45) and (3.46). In particular (3.45) reduces to the loglinear representation

$$\log(m_i^{*(0)}/np_i) = \tau_0 + c_{i1}\tau_1 + c_{i2}\tau_2 + \cdots + c_{is}\tau_s \tag{3.47}$$

in (3.36); and (3.46) is just n times the discrimination information $K(\boldsymbol{\pi}:\mathbf{p})$ defined in (3.34) with $\mathbf{m} = n\boldsymbol{\pi}$.

Setting $\lambda = 1$ and $\mathbf{p} = \mathbf{1}/k$, then (3.45) collapses to the linear or additive model

$$m_i^{*(1)} = \tau_0 + c_{i1}\tau_1 + c_{i2}\tau_2 + \cdots + c_{is}\tau_s, \tag{3.48}$$

and (3.46) is Pearson's X^2 measure of divergence of $\mathbf{m}$ from $n\mathbf{1}/k$

$$X^2 = \sum_{i=1}^{k} \frac{(m_i - n/k)^2}{n/k}.$$

In other words (3.48) minimizes the sum of squared differences between $\mathbf{m}$ and the equiprobable frequency vector $n\mathbf{1}/k$, subject to the constraints (3.44).

In the special case of three-dimensional contingency tables, the principle of generalized minimum power-divergence estimation provides interesting characterizations of two well-known models of no three-factor interaction (Darroch, 1974). Throughout the rest of this section we assume that the vectors $\mathbf{m} = n\boldsymbol{\pi}$, $\mathbf{m}^{*(\lambda)} = n\boldsymbol{\pi}^{*(\lambda)}$, and $\mathbf{p}$ (defined by (3.44) and (3.45)) consist of elements of the form $m_{ijl} = n\pi_{ijl}$, $m_{ijl}^{*(\lambda)} = n\pi_{ijl}^{*(\lambda)}$, and p_{ijl}, where the subscripts range over the three dimensions of the contingency table.

First consider the model of no three-factor interaction given by (3.17). For simplicity we shall rewrite this model as

$$\log(\pi_{ijl}) = \alpha_{ij} + \beta_{il} + \gamma_{jl}, \tag{3.49}$$

for some α_{ij}, β_{il}, and γ_{jl}. We define the constraints in (3.44) to fix the two-dimensional margins π_{ij+}, π_{i+l}, π_{+jl} and define $p_{ijl} = \pi_{i++}\pi_{+j+}\pi_{++l}$ (hence $\mathbf{p}$ is constant for all $\boldsymbol{\pi}$ satisfying the two-dimensional margins) then it is easy to show that (3.49) represents the same model as (3.45) with $\lambda = 0$ (i.e., equation (3.47)). Consequently among all $\boldsymbol{\pi}$ with fixed two-dimensional margins, the no-three-factor-interaction model defines the probability vector that is closest to complete independence (i.e., $\pi_{ijl} = \pi_{i++}\pi_{+j+}\pi_{++l}$; see (3.14)) according to the measure $2I^0(n\boldsymbol{\pi}:n\mathbf{p})$.

An alternative (and different) definition of no three-factor interaction to (3.49) is the Lancaster-additive model (e.g., Darroch and Speed, 1983). For the three-dimensional contingency table, the Lancaster-additive model can be defined as

$$\frac{\pi_{ijl}}{\pi_{i++}\pi_{+j+}\pi_{++l}} = \alpha_{ij} + \beta_{il} + \gamma_{jl}, \tag{3.50}$$

for some α_{ij}, β_{il}, and γ_{jl}. Model (3.50) follows from (3.45) with $\lambda = 1$ by setting the constraints in (3.44) to fix all the two-dimensional margins π_{ij+}, π_{i+l} and π_{+jl} and defining $p_{ijl} = \pi_{i++}\pi_{+j+}\pi_{++l}$ (Section A3 of the Appendix). In other words, among all $\boldsymbol{\pi}$ with fixed two-dimensional margins, the Lancaster-additive model defines the probability vector, which is closest to complete independence according to $2I^1(n\boldsymbol{\pi} : n\mathbf{p})$ (Pearson's X^2).

It is important to realize that unlike the case $\lambda = 0$, the existence of $\mathbf{m}^{*(\lambda)}$ is not always guaranteed for other values of λ. For example, in the case of the Lancaster-additive model, Darroch (1974) provides an example where $\mathbf{m}^{*(1)}$ does not exist. For further discussion and comparison of the Lancaster-additive model and the loglinear model, see Lancaster (1969) and Darroch (1974, 1976).

In Section 8.2 the concept of generalized minimum power-divergence estimation is considered for model selection.

3.6. Model Selection and Testing Strategy for Loglinear Models

For the remainder of this chapter, we shall concentrate our discussion on the loglinear model, which is the model most frequently used in practical applications of categorical data analysis.

Once we have (BAN) estimated the expected cell frequencies $\mathbf{m}$ for the hypothesized loglinear model, we assess the fit of the model by calculating a goodness-of-fit test statistic from the power-divergence family $2I^\lambda(\mathbf{x} : \hat{\mathbf{m}})$ in (3.23). Provided that the calculated value of $2I^\lambda(\mathbf{x} : \hat{\mathbf{m}})$ is not "too large," we conclude that the model adequately explains the data. Three questions that arise here are:

(a) How large is "too large"?
(b) Which value of λ should be chosen for the test statistic?
(c) Which models should be tried?

The question of how large is "too large" is discussed in Chapters 4 and 5. The most commonly used result is to assume that when the hypothesized model is true, the test statistic follows a chi-squared distribution with degrees of freedom given by the number of cells in the table minus the number of parameters estimated, minus one. Therefore "too large" means a value that lies in the upper tail (e.g., the upper 5% tail) of the chi-squared distribution. However this distributional result is only approximate, and the assumptions upon which it is based are not always appropriate. Other sets of assumptions and results are considered in Chapters 4 and 5.

The appropriate choice for λ is also discussed in Chapters 4, 5, and 6; it depends on the type of deviations from the null model we wish to detect. The recommendations of these chapters are summarized briefly herein.

Provided the minimum expected cell frequency is no less than 1, we recommend using $\lambda = 2/3$. The resulting statistic has a distribution that is generally well approximated by the chi-squared distribution and guards reasonably well against the "extreme" alternatives described in Chapter 5. For detecting specific very large or very small ratios of alternative/null expected cell frequencies, we may want to use larger positive or larger negative values of λ, respectively, in accordance with the recommendations of Sections 5.5 and 6.7. However when using values of λ further away from $\lambda = 2/3$, the approximation to the chi-squared distribution is much less appropriate; consequently more accurate approximations to the distribution function are needed as described in Sections 5.1 and 5.2. When the contingency table is large but sparsely filled (i.e., many of the expected frequencies are less than 1) we conjecture that $\lambda = 2/3$ will still be a good choice; however, there are no firm recommendations on the best approximation to use for the critical value. In some cases the chi-squared approximation will still be appropriate, and in other cases the normal approximation of Section 4.3 may be appropriate; this is a topic for future research (Section 8.1).

Which Models Should We Try?

The rest of this section is devoted to discussing the last question: Which models should we try? It is standard practice when fitting loglinear models to require all lower-order effects to be included before a related higher-order effect may be introduced. For example, the interaction term u_{123} may only be included if all of the terms u_{12}, u_{13}, and u_{23} are already present (and in turn these terms require u_1, u_2, u_3, and u all to be present); see (3.13). Such models are called *hierarchical* and are an important class because of the general difficulty of interpreting effects in nonhierarchical models (Fienberg, 1980). Of course there will be specific instances where nonhierarchical models may be appropriate; for example, in modeling a synergistic relationship where a response occurs only when both factors occur together but not alone (Kullback and Keegel, 1984). However unless specific subject knowledge requires the consideration of nonhierarchical models, we recommend using hierarchical models in general.

Even with the restriction to hierarchical models, it can be shown that for a four-dimensional table, there are 113 possible models from which to choose. To provide guidance in selecting a model, there are three popular strategies currently used. To help describe these strategies, we begin by defining the model which includes all interaction terms involving v variables as a uniform model of order v (when $v = 1$, the uniform model consists of simply the main effect for each of the variables in the table, together with the constant term). For example in a four-dimensional table, the hierarchical model including terms u_{12}, u_{13}, u_{14}, u_{23}, u_{24}, and u_{34} is the uniform model of order 2. The three selection strategies are now defined as follows.

Stepwise Selection

This approach is similar to the stepwise procedures used in multiple regression analysis, and consists of three steps.

(a) Initially fit all uniform models of order v for $1 \leq v \leq t$, where t is the dimension of the table;
(b) Find the smallest value of v for which the uniform model of order v fits the data well (and therefore, by assumption, the uniform model of order $v - 1$ fits poorly);
(c) Use a *stepwise* approach (forward selection or backward elimination) or a *best subsets* approach to select the important interaction terms involving v variables, and include these in the model of uniform order $v - 1$.

As a consequence of the further development of models for ordinal data (Agresti, 1984; Goodman, 1984), we recommend that stage (c) be augmented to include interaction terms that reflect any ordering inherent in the categories of the contingency table. As discussed in Section 3.3, ordinal models can result in significantly improved fits with the addition of far fewer parameters than are required by the "standard" interaction terms.

Selection Based on Measures of Marginal and Partial Association

This approach (due to Brown, 1976) is a method of screening terms in order to limit the number that need to be considered in building the model, and is particularly useful for tables with many dimensions. For each term involving v variables ($1 < v < t$), compute the following two statistics.

(a) The *partial association* statistic measures the effect of deleting the given interaction term from the uniform model of order v. For example, in a four-dimensional table, the partial association of u_{12} is measured by comparing the difference of the fit of the hierarchical model containing the terms u_{12}, u_{13}, u_{14}, u_{23}, u_{24}, and u_{34}, with the model in which u_{12} is deleted. This is a conditional test of u_{12} adjusted for all other interactions of the same order.
(b) The *marginal association* statistic measures the effect of the given interaction term from the marginal table obtained by collapsing over all variables not included in the interaction. Consequently the marginal association of u_{12} is measured by calculating the appropriate goodness-of-fit statistic for testing $u_{12} = 0$ from the two-dimensional subtable for variables 1 and 2. This is an unconditional test of u_{12} ignoring all other variables and their interactions.

If both the partial and marginal association tests are significant, the given term should be included in the model; if both tests are insignificant, the term should not be included. If one test is significant while the other is not, the term should be investigated further. Due to the multiplicity of tests being performed here, it is important to realize that the significance levels are only a guide to the relative importance of terms, and should not be considered formal tests of significance.

Selection Based on Standardized Parameter Estimates

The third approach is to estimate all the possible terms in the saturated model, and compare their standardized estimates (e.g., Bishop et al., 1975, Chapter 4). The terms with the largest standardized estimates should be included in the model.

In conclusion, it is important to realize that no strategy can be guaranteed to select the "correct" model, and different selection algorithms can yield different models. Clearly the researcher should use all available prior information on the data set and knowledge about the problem and objectives of the study in order to derive a parsimonious model with some useful physical interpretation. Further discussion and applications of model selection algorithms can be found in Benedetti and Brown (1978), Bishop et al. (1975), Dillon and Goldstein (1984) and Freeman (1987).

CHAPTER 4

Testing the Models: Large-Sample Results

Chapter 3 presented the terminology and concepts for loglinear model fitting. In the present chapter, the emphasis changes from defining models to testing the model fit using the power-divergence goodness-of-fit statistic.

In Chapters 2 and 3, we saw that larger values of the goodness-of-fit statistic indicate a greater discrepancy between the data and the model. If we decide to reject the model, how certain can we be that we have made the right decision (how large must be statistic be)? Alternatively, if we decide to accept the model, what is the chance that, in fact, the model is false?

From the perspective of the statistical literature, these two questions address, respectively: (a) calculating the $\alpha 100\%$ *critical value* for the test statistic (i.e., the value above which we expect only an $\alpha 100\%$ chance of an observation when the hypothesized model is true); and (b) assessing the *efficiency* of the test statistic. The first two sections of this chapter answer these questions for large samples under the assumptions most commonly considered when applying goodness-of-fit tests, namely, increasing sample size but fixed cell number. We refer to these as the *classical (fixed-cells) assumptions*. There are, however, many practical situations in which this set of assumptions is inappropriate, and in Section 4.3 we consider an alternative set, which we call *sparseness assumptions*. Section 4.4 compares the large-sample performance of the power-divergence family from the results of Sections 4.1, 4.2, and 4.3. Finally we offer guidelines for choosing a test statistic in Section 4.5.

4.1. Significance Levels under the Classical (Fixed-Cells) Assumptions

The comparative survey of Pearson's X^2 ($\lambda = 1$) and the loglikelihood ratio statistic G^2 ($\lambda = 0$), contained at the end of this book in the Historical Perspective, indicates that generally the chi-squared distribution is used to quantify the goodness-of-fit of a hypothesized null model. In particular, for a model with k cells and s estimated parameters, we approximate $Pr(X^2 \geq c_\alpha)$ (or $Pr(G^2 \geq c_\alpha)$) by $Pr(\chi^2_{k-s-1} \geq c_\alpha)$, where c_α is the *critical value* of the test and χ^2_{k-s-1} is a chi-squared random variable with $k - s - 1$ degrees of freedom. We choose the critical value c_α so that $Pr(\chi^2_{k-s-1} \geq c_\alpha) = \alpha$ for some small probability α (e.g., $\alpha = 0.05$) and denote c_α by $\chi^2_{k-s-1}(\alpha)$; we then reject the hypothesized model if the value of X^2 (or G^2) is greater than or equal to $\chi^2_{k-s-1}(\alpha)$ (Section 2.2). The α-value used is called the *significance level* of the test, and keeps the probability of rejecting H_0, given H_0 is true, below α.

This approach to testing goodness of fit raises two important questions: (a) Upon what asymptotic (i.e., large-sample) assumptions does the approximation to the chi-squared distribution depend; and (b) can we extend this result to the power-divergence statistic, defined from (2.16) as

$$2nI^\lambda(\mathbf{X}/n : \boldsymbol{\pi}) = \frac{2}{\lambda(\lambda+1)} \sum_{i=1}^{k} X_i \left[\left(\frac{X_i}{n\pi_i} \right)^\lambda - 1 \right]; \qquad -\infty < \lambda < \infty? \quad (4.1)$$

The rest of this section concentrates on answering these questions. Throughout, we shall assume $\mathbf{X} = (X_1, X_2, \ldots, X_k)$ is a multinomial random vector from which we observe $\mathbf{x} = (x_1, x_2, \ldots, x_k)$; see (2.4) for the definition. We write $\mathbf{X}$ is multinomial $\mathrm{Mult}_k(n, \boldsymbol{\pi})$, where n is the total number of counts over the k cells, and $\boldsymbol{\pi} = (\pi_1, \pi_2, \ldots, \pi_k)$ is the true (unknown) probability vector for observing a count in each cell. The results derived here also hold true for the Poisson and product-multinomial sampling models discussed in Chapter 3, under certain restrictions on the hypothesized model (Bishop et al., 1975, chapter 3). Equivalence of these three sampling models is discussed in more detail by Bishop et al. (1975, chapter 13).

The Simple Hypothesis—No Parameter Estimation

Consider first the case of a simple null hypothesis for $\boldsymbol{\pi}$:

$$H_0: \boldsymbol{\pi} = \boldsymbol{\pi}_0, \quad (4.2)$$

where $\boldsymbol{\pi}_0 = (\pi_{01}, \pi_{02}, \ldots, \pi_{0k})$ is completely specified, and each $\pi_{0i} > 0$. To derive the asymptotic chi-squared distribution for Pearson's X^2 statistic ($\lambda = 1$) assuming H_0, we use the following basic results (Section A4 of the Appendix):

(i) Assume $\mathbf{X}$ is a random vector with a multinomial distribution $\text{Mult}_k(n, \boldsymbol{\pi})$, and H_0: $\boldsymbol{\pi} = \boldsymbol{\pi}_0$ from (4.2). Then $\sqrt{n}(\mathbf{X}/n - \boldsymbol{\pi}_0)$ converges in distribution to a multivariate normal random vector as $n \to \infty$.

(ii)
$$X^2 = \sum_{i=1}^{k} \frac{(X_i - n\pi_{0i})^2}{n\pi_{0i}}$$

can be written as a quadratic form in $\sqrt{n}(\mathbf{X}/n - \boldsymbol{\pi}_0)$, and X^2 converges in distribution (as $n \to \infty$) to the quadratic form of the multivariate normal random vector in (i).

(iii) Certain quadratic forms of multivariate normal random vectors are distributed as chi-squared random variables. In the case of (i) and (ii), X^2 converges in distribution to a central chi-squared random variable with $k - 1$ degrees of freedom.

Using results (i), (ii), and (iii), we deduce that

$$Pr(X^2 \geq c) \to Pr(\chi^2_{k-1} \geq c), \qquad \text{as } n \to \infty,$$

for any $c \geq 0$, and in particular,

$$Pr(X^2 \geq \chi^2_{k-1}(\alpha)) \to \alpha, \qquad \text{as } n \to \infty. \tag{4.3}$$

From results (i), (ii), and (iii) we see that using the chi-squared distribution to calculate the critical value of Pearson's X^2 test for an $\alpha 100\%$ significance level assumes: (a) the sample size n is large enough for (i), and hence (4.3), to be approximately true; and (b) the number of cells is fixed, so that k is small relative to n. Condition (b) implies that each expected cell frequency $n\pi_{0i}$ should be large, since $n\pi_{0i} \to \infty$, as $n \to \infty$, for each $i = 1, \ldots, k$.

The adequacy of (4.3) for small values of n and small expected cell frequencies is discussed in Chapter 5. The rest of this section is devoted to extending (4.3) to other members of the power-divergence family and to presenting a parallel result for models containing unknown parameters, estimated using the minimum power-divergence estimate defined in (3.25).

The extension of (4.3) from Pearson's X^2 ($\lambda = 1$) to the general λ of the power-divergence family (4.1), uses a Taylor-series expansion. Observe that (4.1) can be rewritten as

$$2nI^\lambda(\mathbf{X}/n : \boldsymbol{\pi}_0) = \frac{2n}{\lambda(\lambda + 1)} \sum_{i=1}^{k} \pi_{0i}\left[\left(1 + \frac{X_i - n\pi_{0i}}{n\pi_{0i}}\right)^{\lambda+1} - 1\right]$$

(provided $\lambda \neq 0$, $\lambda \neq -1$); now set $V_i = (X_i - n\pi_{0i})/n\pi_{0i}$ and expand in a Taylor series for each λ, giving:

$$\begin{aligned} 2nI^\lambda(\mathbf{X}/n : \boldsymbol{\pi}_0) &= \frac{2n}{\lambda(\lambda + 1)} \sum_{i=1}^{k} \pi_{0i}\left[(\lambda + 1)V_i + \frac{\lambda(\lambda + 1)}{2} V_i^2 + o_p(1/n)\right] \\ &= n\left[\sum_{i=1}^{k} \pi_{0i} V_i^2 + o_p(1/n)\right], \end{aligned} \tag{4.4}$$

where $o_p(1/n)$ (which here depends on λ) represents a stochastic term that converges to 0 in probability faster than $1/n$ as $n \to \infty$. An identical result to (4.4) follows for the special cases $\lambda = 0$ and $\lambda = -1$ by expanding the limiting forms of $2nI^\lambda(\mathbf{X}/n : \boldsymbol{\pi}_0)$ given in Section 2.4. The justification of the $o_p(1/n)$ term comes from result (i), given earlier, because $\sqrt{n\pi_{0i}}V_i$ is asymptotically normally distributed, as $n \to \infty$; $i = 1, \ldots, k$. Further details on stochastic convergence can be found in Bishop et al. (1975, pp. 475–484). Finally, we rewrite (4.4) for each λ, as

$$2nI^\lambda(\mathbf{X}/n : \boldsymbol{\pi}_0) = 2nI^1(\mathbf{X}/n : \boldsymbol{\pi}_0) + o_p(1); \qquad -\infty < \lambda < \infty,$$

by noting that the first term on the right-hand side of (4.4) is simply Pearson's X^2. This result is sufficient for us to conclude that each member of the power-divergence family has the same asymptotic distribution as Pearson's X^2 statistic, so that

$$Pr(2nI^\lambda(\mathbf{X}/n : \boldsymbol{\pi}_0) \geq \chi^2_{k-1}(\alpha)) \to \alpha, \qquad \text{as } n \to \infty; \tag{4.5}$$

for each $\lambda \in (-\infty, \infty)$ and each $\alpha \in (0, 1)$.

Both (4.3) and (4.4) rely on results (i), (ii), and (iii). Therefore the asymptotic equivalence of the power-divergence family of statistics relies on the number of cells k being fixed, and the expected cell frequencies $n\pi_{0i}$ being large, for all $i = 1, \ldots, k$. In Section 4.3 we see that without these conditions, the statistics are no longer asymptotically equivalent.

An Example: Greyhound Racing

As an illustration, we consider testing a model to predict the probability of winning at a greyhound race track in Australia (Read and Cowan, 1976). Data collected on 595 races give the starting numbers of the eight dogs included in each race ordered according to the race finishing positions (the starting numbers are always the digits $1, \ldots, 8$; 1 denotes the dog started on the fence, 2 denotes second from the fence, etc.). For example, the result 4, 3, 7, 8, 1, 5, 2, 6 would indicate dog 4 came in first, dog 3 was second, ..., and dog 6 was last. We assume throughout that the starting numbers are randomly assigned at the beginning of each race. The simplest model to hypothesize is that all race results are equally likely; but since there are $8! = 40{,}320$ possible race results, this model cannot be tested with any confidence using a sample of only 595 race results. Instead, we group the results into eight cells according to which starting number comes in first. Now we test the hypothesis that all starting numbers have an equal chance of coming in first regardless of the positions of the other seven dogs; that is,

$$H_0: \pi_i = 1/8; \qquad i = 1, \ldots, 8$$

where $\pi_i = Pr(\text{dog number } i \text{ wins})$. Table 4.1 gives the observed and expected number of races which fall into each of the eight cells.

Table 4.1. Observed and Equiprobable Expected Frequencies for Dog i to Win Regardless of the Other Finishing Positions

Dog i	Observed	Expected
1	104	74.375
2	95	74.375
3	66	74.375
4	63	74.375
5	62	74.375
6	58	74.375
7	60	74.375
8	87	74.375
Total	595	595.0

Source: Read and Cowan (1976).

Table 4.2. Computed Values of the Power-Divergence Statistic (4.1) for the Data in Table 4.1

λ									
-5	-2	-1	$-1/2$	0	1/2	2/3	1	2	5
28.73	28.40	28.87	29.22	29.65	30.17	30.37	30.79	32.32	40.03

To test the null hypothesis that all winning numbers are equally likely, various members of the power-divergence family (4.1) are given in Table 4.2.

Using result (4.5), we test the equiprobable null hypothesis at the approximate 5% significance level by comparing the computed values of Table 4.2 against the 95th percentile of a chi-squared distribution with $8 - 1 = 7$ degrees of freedom; that is, $\chi_7^2(0.05) = 14.07$. Since all the computed values are substantially larger than 14.07, we conclude it is extremely unlikely that all starting numbers have an equal chance of finishing first. We shall return to this example later with a revised model.

The General Hypothesis and Parameter Estimation

In Chapter 3, many of the hypotheses discussed require unspecified parameters to be estimated before calculating the goodness-of-fit statistics. More specifically, we rewrite (3.24) as

$$H_0: \boldsymbol{\pi} \in \Pi_0, \tag{4.6}$$

where Π_0 is a set of possible values for $\boldsymbol{\pi}$. For example, in the case of testing independence for a two-dimensional contingency table $\Pi_0 = \{\pi_{ij}: \pi_{ij} > 0; \sum_{i=1}^r \sum_{j=1}^c \pi_{ij} = 1; \pi_{ij} = \pi_{i+}\pi_{+j}\}$, since each probability π_{ij} must satisfy (3.1) and lie inside the $(k-1)$-dimensional simplex (that is, each $\pi_{ij} > 0$ and $\sum_{i=1}^r \sum_{j=1}^c \pi_{ij} = 1$). We must choose (estimate) one value $\hat{\boldsymbol{\pi}} \in \Pi_0$ that is "most consistent" with the observed proportions $\mathbf{x}/n$, and then test H_0 by calculating $2nI^\lambda(\mathbf{x}/n : \hat{\boldsymbol{\pi}})$ for some fixed $-\infty < \lambda < \infty$. The most sensible way to estimate $\boldsymbol{\pi}$ is to choose the $\hat{\boldsymbol{\pi}} \in \Pi_0$ that is closest to $\mathbf{x}/n$ with respect to the measure $2nI^\lambda(\mathbf{x}/n : \boldsymbol{\pi})$. This leads to the minimum power-divergence estimate defined in (3.25), namely, the value $\hat{\boldsymbol{\pi}}^{(\lambda)}$ which satisfies

$$I^\lambda(\mathbf{x}/n : \hat{\boldsymbol{\pi}}^{(\lambda)}) = \inf_{\boldsymbol{\pi} \in \Pi_0} I^\lambda(\mathbf{x}/n : \boldsymbol{\pi}); \qquad -\infty < \lambda < \infty. \tag{4.7}$$

For the example of testing independence for a two-dimensional contingency table, we saw in Section 3.4 that for $\lambda = 0$, (4.7) defines the maximum likelihood estimates (MLEs) $\hat{\pi}_{ij}^{(0)} = x_{i+}x_{+j}/n^2$; $i = 1, \ldots, r$; $j = 1, \ldots, c$.

How does the estimation of parameters affect results (4.3), (4.4), and (4.5)? Do we need to place any restrictions on the set Π_0 (other than requiring it to be a subset of the $(k-1)$-dimensional simplex)? The now-classical result for X^2 ($\lambda = 1$) and G^2 ($\lambda = 0$) quoted at the beginning of this section indicates that we may still use the chi-squared distribution for large sample size n, provided we subtract one degree of freedom for each parameter estimated. We now examine briefly the conditions necessary for this procedure to yield the correct asymptotic distribution and extend the result to all members of the power-divergence family.

To understand the restrictions we must place on the set Π_0 it is helpful to define $\boldsymbol{\pi}$ as a function of s parameters $\boldsymbol{\theta} = (\theta_1, \theta_2, \ldots, \theta_s)$, that is, $\pi_i = f_i(\boldsymbol{\theta})$; $i = 1, \ldots, k$, or

$$\boldsymbol{\pi} = \mathbf{f}(\boldsymbol{\theta}), \tag{4.8}$$

where the vector of parameters $\boldsymbol{\theta}$ ranges over a subset of R^s. It will be seen later that this subset needs to satisfy certain regularity conditions.

For example, testing independence in a two-dimensional contingency table with $r = c = 2$ results in two parameters, $\boldsymbol{\theta} = (\theta_1, \theta_2) \in [0, 1] \times [0, 1]$. The reason is clear if we recall that the linear constraints on the probabilities require $\pi_{2+} = 1 - \pi_{1+}$ and $\pi_{+2} = 1 - \pi_{+1}$. Therefore, writing $\theta_1 = \pi_{1+}$ and $\theta_2 = \pi_{+1}$ we can define $\mathbf{f}(\boldsymbol{\theta})$ in (4.8) to be $\pi_{11} = f_{11}(\theta_1, \theta_2) = \theta_1\theta_2$; $\pi_{12} = f_{12}(\theta_1, \theta_2) = \theta_1(1 - \theta_2)$; $\pi_{21} = f_{21}(\theta_1, \theta_2) = (1 - \theta_1)\theta_2$; and $\pi_{22} = f_{22}(\theta_1, \theta_2) = (1 - \theta_1)(1 - \theta_2)$. The MLE of $\boldsymbol{\theta} = (\theta_1, \theta_2)$ is $\hat{\boldsymbol{\theta}} = (\hat{\theta}_1, \hat{\theta}_2) = (x_{1+}/n, x_{+1}/n)$, as we observed already in Chapter 3.

Birch's Regularity Conditions

Birch (1964) defines a set of six *regularity conditions* that are sufficient to ensure that any minimum power-divergence estimate (4.7) will be best asymptotically normal (BAN) (Section A5 of the Appendix). BAN estimators (i.e., estimates considered as random vectors by replacing the observed vector $\mathbf{x}$ with the random vector $\mathbf{X}$) have three important properties:

(a) They are *consistent*, i.e., the estimator converges to the true value of the estimated parameter as $n \to \infty$.
(b) They are *asymptotically normally distributed*. We rely on this result to derive the asymptotic chi-squared distribution for $2nI^{\lambda}(\mathbf{X}/n : \hat{\boldsymbol{\pi}})$.
(c) They are asymptotically *efficient*, in that no other estimator can have smaller variance, as $n \to \infty$.

The six regularity conditions of Birch (1964) (described in Section A5 of the Appendix) require that $\mathbf{f}$ satisfies certain smoothness properties and that $\pi_i = f_i(\boldsymbol{\theta})$ is positive for all i (ensuring that there are no fewer than k cells). Finally, the conditions ensure that the model really does have s parameters and not fewer (Bishop et al., 1975, p. 510 provide further details). These conditions are satisfied by all the loglinear models of Chapter 3.

Using the asymptotic normality of the BAN estimator $\hat{\boldsymbol{\pi}}$, it is straightforward to parallel results (i), (ii), and (iii), given earlier, for the general hypothesis (4.6) (Section A6 of the Appendix), giving:

(i*) Assume $\mathbf{X}$ is a random vector with a multinomial distribution $\text{Mult}_k(n, \boldsymbol{\pi})$, and $H_0: \boldsymbol{\pi} = \mathbf{f}(\boldsymbol{\theta}^*) \in \Pi_0$, for some (unknown) $\boldsymbol{\theta}^* = (\theta_1^*, \theta_2^*, \ldots, \theta_s^*) \in \Theta_0 \subset R^s$. Provided $\mathbf{f}$ satisfies Birch's regularity conditions and $\hat{\boldsymbol{\pi}} \in \Pi_0$ is a BAN estimator of $\mathbf{f}(\boldsymbol{\theta}^*)$, then $\sqrt{n}(\mathbf{X}/n - \hat{\boldsymbol{\pi}})$ converges in distribution to a multivariate normal random vector as $n \to \infty$.

(ii*)
$$X^2 = \sum_{i=1}^{k} \frac{(X_i - n\hat{\pi}_i)^2}{n\hat{\pi}_i}$$

can be written as a quadratic form in $\sqrt{n}(\mathbf{X}/n - \hat{\boldsymbol{\pi}})$, and X^2 converges in distribution (as $n \to \infty$) to the quadratic form of the multivariate normal random vector in (i*).

(iii*) Certain quadratic forms of multivariate normal random vectors are distributed as chi-squared random variables. In the case of (i*) and (ii*), X^2 converges in distribution to a central chi-squared random variable with $k - s - 1$ degrees of freedom.

A similar Taylor-series expansion to (4.4) (see Section A6 of the Appendix) allows us to conclude from (i*), (ii*), and (iii*) that for the models in Chapter 3,

$$Pr(2nI^{\lambda}(\mathbf{X}/n : \hat{\boldsymbol{\pi}}) \geq c) \to Pr(\chi^2_{k-s-1} \geq c), \qquad \text{as } n \to \infty;$$

for each $\lambda \in (-\infty, \infty)$, $c \geq 0$. Thus for $c = \chi^2_{k-s-1}(\alpha)$, $\alpha \in (0, 1)$,

$$Pr(2nI^{\lambda}(\mathbf{X}/n : \hat{\boldsymbol{\pi}}) \geq \chi^2_{k-s-1}(\alpha)) \to \alpha, \qquad \text{as } n \to \infty. \tag{4.9}$$

In particular, if $\hat{\boldsymbol{\pi}}^{(\lambda)} \in \Pi_0$ is the minimum power-divergence estimator of $\boldsymbol{\pi}$, then for each $\lambda \in (-\infty, \infty)$ and each $\alpha \in (0, 1)$,

$$Pr(2nI^{\lambda}(\mathbf{X}/n : \hat{\boldsymbol{\pi}}^{(\lambda)}) \geq \chi^2_{k-s-1}(\alpha)) \to \alpha, \qquad \text{as } n \to \infty. \tag{4.10}$$

For example, consider again the test of independence in a two-dimensional contingency table where the number of cells $k = rc$. For the case $r = c = 2$ we have just seen that $s = 2$ which gives $k - s - 1 = 4 - 2 - 1 = 1$ (for general r and c, $s = (r - 1) + (c - 1)$, giving $k - s - 1 = (r - 1)(c - 1)$). Therefore (4.9) indicates that the power-divergence statistic for testing independence in 2×2 contingency tables is approximately chi-squared with one degree of freedom when the null model is true (a very well-known result in the cases $\lambda = 0$ and $\lambda = 1$).

The Greyhound Example Revisited

As a further illustration, we return briefly to the greyhound racing data of Table 4.1. From Table 4.2 we saw that the equiprobable hypothesis does not fit these data well. Now we consider a slightly more complicated model, which involves both first- and second-place getters. Let $\pi_{ij} = Pr$(dog i finishes first and dog j finishes second), then if we assume that dog i finishes first with probability $\pi_i = \pi_{i+}$, we may consider the ensuing race for second position to be a *subrace* of the seven remaining dogs. In other words, the model becomes

$$\begin{aligned} H_0\colon \pi_{ij} &= Pr(\text{dog } i \text{ wins}) \cdot Pr(\text{dog } j \text{ wins among remaining dogs}) \\ &= Pr(\text{dog } i \text{ wins}) \cdot Pr(\text{dog } j \text{ wins given dog } i \text{ is not in the race}) \\ &= \pi_i \pi_j/(1 - \pi_i) \end{aligned}$$

for $i = 1, \ldots, 8$; $j = 1, \ldots, 8$; $i \neq j$. Of course $\pi_{ii} = 0$ for $i = 1, \ldots, 8$.

The subrace null model has seven parameters to be estimated, namely, $\boldsymbol{\theta} = (\theta_1, \theta_2, \ldots, \theta_7) = (\pi_1, \pi_2, \ldots, \pi_7) \in \{\boldsymbol{\theta}\colon \boldsymbol{\theta} \in (0, 1)^7 \text{ and } \theta_1 + \theta_2 + \cdots + \theta_7 < 1\}$, for which we define $\mathbf{f}(\boldsymbol{\theta})$ in (4.8) to be

$$\begin{aligned} \pi_{ij} = f_{ij}(\boldsymbol{\theta}) &= \frac{\theta_i \theta_j}{1 - \theta_i}; & i &= 1, \ldots, 7; j = 1, \ldots, 7; i \neq j \\ &= \frac{\theta_i(1 - \theta_1 - \cdots - \theta_7)}{1 - \theta_i}; & i &= 1, \ldots, 7; j = 8 \\ &= \frac{(1 - \theta_1 - \cdots - \theta_7)\theta_j}{\theta_1 + \cdots + \theta_7}; & i &= 8; j = 1, \ldots, 7. \end{aligned}$$

The MLE $\hat{\boldsymbol{\theta}}$ of $\boldsymbol{\theta}$ requires an iterative solution which Read and Cowan (1976) implement to give $\hat{\boldsymbol{\theta}} = (0.1787, 0.1360, 0.1145, 0.1117, 0.1099, 0.1029, 0.1122)$. Table 4.3 contains the observed and expected frequencies for first- and second-place getters under the subrace null model.

Table 4.4 contains the computed values of the power-divergence statistic (4.1) for the data in Table 4.3. Result (4.9) indicates that we should reject the null model at the 5% significance level if the value of $2nI^{\lambda}(\mathbf{x}/n : \hat{\boldsymbol{\pi}})$ is greater than or equal to $\chi^2_{56-7-1}(0.05) = 64.1$. From Table 4.4 we see that for $\lambda \in (-1, 2]$ (the recommended interval in Section 4.5 from which to choose λ) all statistic values are less than the critical value. For λ outside the interval $(-1, 2]$, the $\chi^2_{48}(\alpha)$ value tends to be too small for the exact $\alpha 100\%$ significance level (Section 5.3) and the test statistic tends to be very sensitive to single cell depar-

Table 4.3. Observed and Expected Frequencies for the First Two Place-Getters Assuming the Subrace Model

	Second								
First	1	2	3	4	5	6	7	8	Total
1*		14	11	11	17	17	15	19	104
		17.6	14.8	14.5	14.2	13.3	14.5	17.4	106.3
2	22		12	14	15	6	14	12	95
	16.7		10.7	10.5	10.3	9.6	10.5	12.5	80.9
3	13	10		9	9	12	8	5	66
	13.8	10.5		8.6	8.5	7.9	8.6	10.3	68.1
4	10	10	5		13	12	8	5	63
	13.4	10.2	8.6		8.2	7.7	8.4	10.0	66.5
5	12	7	8	9		7	9	10	62
	13.1	10.0	8.4	8.2		7.6	8.2	9.8	65.4
6	10	8	10	5	7		9	9	58
	12.2	9.3	7.8	7.6	7.5		7.7	9.1	61.2
7	8	6	12	8	6	9		11	60
	13.4	10.2	8.6	8.4	8.3	7.7		10.1	66.8
8	27	9	14	16	4	4	13		87
	16.5	12.5	10.5	10.3	10.1	9.5	10.3		79.7
Total	102	64	72	72	71	67	76	71	595
	99.1	80.3	69.5	68.0	67.1	63.4	68.3	79.3	595.0

* First row of each pair of rows gives observed and second row gives expected frequencies.
Source: Read and Cowan (1976).

Table 4.4. Computed Values of the Power-Divergence Statistic (4.1) for the Data in Table 4.3

λ									
-5	-2	-1	$-1/2$	0	1/2	2/3	1	2	5
145.74	69.82	61.91	59.46	57.79	56.81	56.62	56.43	57.33	73.67

tures (Chapter 6). Therefore with no specific alternative models in mind we recommend using $\lambda = 2/3$ (Section 4.5): $2nI^{2/3}(\mathbf{x}/n : \hat{\boldsymbol{\pi}}) = 56.6 < \chi^2_{48}(0.05) = 64.1$, and hence we should accept the null subrace model.

A Summary of the Assumptions Necessary for the Chi-Squared Results (4.9) and (4.10)

The derivation of (4.9) and (4.10) shows us that the main assumptions for these results are: (a) the hypothesis H_0 given by (4.6) is true; (b) the sample size n is large; (c) the number of cells k is small relative to n (and that all the expected cell frequencies $n\pi$ are large); (d) unknown parameters are estimated with BAN estimates; and (e) the models satisfy the regularity conditions of Birch (1964) (which is true of all the models we consider).

Condition (a) is discussed in Section 4.2, where we assume the null hypothesis given by (4.6) is false, and we compare distributions under alternative models to H_0. Condition (b) is evaluated in detail in Chapter 5 where we assess the adequacy of the chi-squared distribution for small values of n. Condition (c) is discussed in Section 4.3 under the *sparseness* assumptions where n and k are of similar magnitude (e.g., contingency tables with many cells but relatively few observed counts). We close this section with a well-known example where condition (d) is not satisfied.

The Chernoff-Lehmann Statistic

Consider testing the hypothesis that a set of observations $y_1, y_2, \ldots, y_n$ comes from a parametric family of distributions given by $G(y; \boldsymbol{\theta})$ where $\boldsymbol{\theta} = (\theta_1, \theta_2, \ldots, \theta_s)$ must be estimated from the sample. For example: $G(y; \boldsymbol{\theta})$ might be the normal distribution with $\boldsymbol{\theta} = (\mu, \sigma)$ where μ is the mean and σ is the standard deviation of the distribution. To perform a multinomial goodness-of-fit test, the observed $\{y_j\}$ are grouped into k cells with frequency vector $\mathbf{x} = (x_1, x_2, \ldots, x_k)$ where $x_i = \#\{y_j\text{s in cell } i\}$; $i = 1, \ldots, k$.

If we estimate $\boldsymbol{\theta}$ using maximum likelihood (i.e., the minimum power-divergence estimate with $\lambda = 0$) based on $\mathbf{x}$, then the results of this section will hold; however we could estimate $\boldsymbol{\theta}$ directly from the ungrouped $\{y_j\}$. For example, in the case of a normal distribution we could use $\hat{\mu} = \bar{y} = \sum_{j=1}^{n} y_j/n$, and $\hat{\sigma} = [\sum_{j=1}^{n} (y_j - \bar{y})^2/(n-1)]^{1/2}$. In general such estimates will not satisfy the BAN properties, and the resulting power-divergence statistic will be stochastically larger than a chi-squared random variable with $k - s - 1$ degrees of freedom. In the special case $\lambda = 1$, the asymptotic distribution of this statistic has been derived by Chernoff and Lehmann (1954), and is called the Chernoff-Lehmann statistic (Section 4 of the Historical Perspective). This same asymptotic distribution is shown to hold for $-\infty < \lambda < \infty$ in Section A7 of the Appendix. When the number of cells is chosen to be of the same order of magnitude as the sample size, Gan (1985) derives the asymptotic distribution of X^2 to be normal provided θ is a one-dimensional location parameter estimated by the ungrouped sample median (Sections 4.3 and 8.1).

4.2. Efficiency under the Classical (Fixed-Cells) Assumptions

So far we have compared the large-sample distributions of the power-divergence family when the hypothesized model (4.2) or (4.6) is true. An acceptance of this model based on the power-divergence test statistic (for some fixed

$-\infty < \lambda < \infty$) is accompanied by some possibility of having made an incorrect decision. The question we must ask now is: How efficient is the test statistic at distinguishing the hypothesized null model from some alternative "true" model for the sampled population?

Consider the general null model (4.6) together with a hypothesized alternative model:

$$H_0\colon \boldsymbol{\pi} \in \Pi_0$$

versus (4.11)

$$H_1\colon \boldsymbol{\pi} \in \Pi_1.$$

In Section 4.1, we discussed using the chi-squared distribution to find a critical value c_α, to keep the probability of rejecting H_0, given H_0 is true, at or below a prespecified level α; we look for the smallest c_α that satisfies

$$Pr(2nI^\lambda(\mathbf{X}/n : \hat{\boldsymbol{\pi}}) \geq c_\alpha | \boldsymbol{\pi} \in \Pi_0) \leq \alpha.$$

We then carry out the test of the null model by rejecting H_0 if $2nI^\lambda(\mathbf{x}/n : \hat{\boldsymbol{\pi}})$ is observed to be greater than or equal to c_α. Otherwise H_0 is accepted, indicating that we have no significant evidence to refute the null model. To quantify the chance of accepting an erroneous model, we define the power of the test at $\boldsymbol{\pi}$ to be

$$\beta^\lambda(\boldsymbol{\pi}) = Pr(2nI^\lambda(\mathbf{X}/n : \hat{\boldsymbol{\pi}}) \geq c_\alpha | \boldsymbol{\pi} \in \Pi_1). \tag{4.12}$$

The closer the power is to 1, the better the test.

In the spirit of Section 4.1, it would be sensible to compare the large-sample values of $\beta^\lambda(\boldsymbol{\pi})$ for different members of the power-divergence family of statistics. However, if Π_1 from (4.11) is a set of nonlocal probability vectors that are a fixed distance away from the vectors in Π_0, then $\beta^\lambda(\boldsymbol{\pi})$ tends to 1 as the sample size increases. In other words, the power-divergence test is consistent.

Pitman Asymptotic Relative Efficiency

To produce some less trivial asymptotic powers that are not all equal to 1, Cochran (1952) describes using a set of local alternative hypotheses that converge as n increases. In particular, consider

$$H_{1,n}\colon \boldsymbol{\pi} = \boldsymbol{\pi}^* + \boldsymbol{\delta}/\sqrt{n}, \tag{4.13}$$

where $\boldsymbol{\pi}^* \in \Pi_0$ is the true (but in general unknown) value of $\boldsymbol{\pi}$ under $H_0\colon \boldsymbol{\pi} \in \Pi_0$, and $\boldsymbol{\delta} = (\delta_1, \delta_2, \ldots, \delta_k)$; $\sum_{i=1}^k \delta_i = 0$. Because of the convergence of $H_{1,n}$ to H_0, it is possible to use some extensions of the results in Section 4.1 and generalize a result due to Mitra (1958) to give

$$\begin{aligned} \beta^\lambda(\boldsymbol{\pi}^* + \boldsymbol{\delta}/\sqrt{n}) &= Pr(2nI^\lambda(\mathbf{X}/n : \hat{\boldsymbol{\pi}}) \geq c_\alpha | \boldsymbol{\pi} = \boldsymbol{\pi}^* + \boldsymbol{\delta}/\sqrt{n}) \\ &\to Pr(\chi(\gamma)^2_{k-s-1} \geq c_\alpha), \qquad \text{as } n \to \infty;\ -\infty < \lambda < \infty. \end{aligned} \tag{4.14}$$

Here $\chi(\gamma)^2_{k-s-1}$ represents a noncentral chi-squared random variable with $k - s - 1$ degrees of freedom (recall that k is the number of cells and s is the number of BAN-estimated parameters) and noncentrality parameter $\gamma = \sum_{i=1}^{k} \delta_i^2/\pi_i^*$. The details of the proof are given in Section A8 of the Appendix. The result (4.14) does not depend on λ, and this, together with (4.9), indicates that the members of the power-divergence family are asymptotically equivalent under both H_0: $\boldsymbol{\pi} \in \Pi_0$ and local alternatives (4.13).

The *Pitman asymptotic relative efficiency* (a.r.e.) of $2nI^{\lambda_1}(\mathbf{X}/n : \hat{\boldsymbol{\pi}})$ to $2nI^{\lambda_2}(\mathbf{X}/n : \hat{\boldsymbol{\pi}})$ is defined to be the limiting ratio of the sample sizes required for the two tests to maintain the same prespecified power for a given significance level, as $n \to \infty$. In other words, we assume that for a given sample size n and asymptotic $\alpha 100\%$ significance level, there exists a number N_n such that

$$\beta^{\lambda_1}(\boldsymbol{\pi}^* + \boldsymbol{\delta}/\sqrt{n}) = \beta^{\lambda_2}(\boldsymbol{\pi}^* + \boldsymbol{\delta}/\sqrt{N_n}) \to \beta < 1, \qquad \text{as } n \to \infty,$$

and $N_n \to \infty$ as $n \to \infty$; where β^{λ_1} and β^{λ_2} are the asymptotic power functions for $\lambda = \lambda_1$ and λ_2, respectively, and β is some prespecified value. Then the limiting ratio of n/N_n is the Pitman a.r.e. of $2nI^{\lambda_1}(\mathbf{X}/n : \hat{\boldsymbol{\pi}})$ to $2nI^{\lambda_2}(\mathbf{X}/n : \hat{\boldsymbol{\pi}})$ (e.g., Rao, 1973, pp. 467–470; Wieand, 1976, has a more general definition of Pitman a.r.e.). From (4.14) the asymptotic power functions β^{λ_1} and β^{λ_2} will be equal, which means the Pitman a.r.e. will be 1 for any two values λ_1 and λ_2. Consequently we cannot optimize our choice of λ based on the Pitman definition of efficiency.

Efficiency for Nonlocal Alternatives

Using local alternatives that converge to the null model is only one way of ensuring that the power of a consistent test is bounded away from 1 in large samples. Another method is to use a nonlocal (fixed) alternative model but make the significance level, $\alpha 100\%$, decrease steadily with n (e.g., Cochran, 1952; Hoeffding, 1965). This approach is similar to the one introduced by Bahadur (1960, 1971), which he calls *stochastic comparison* but is now commonly known as *Bahadur efficiency*. Cressie and Read (1984) outline the Bahadur efficiency for the power-divergence statistic, from which it is shown that no member of the power-divergence family can be more Bahadur efficient than the loglikelihood ratio statistic G^2 ($\lambda = 0$). The results on Bahadur efficiency are proved only for hypothesized models requiring no parameter estimation.

Yet another approach is pursued by Broffitt and Randles (1977) (and corrected by Selivanov, 1984), who show that for a nonlocal (fixed) alternative (and a simple null hypothesis H_0: $\boldsymbol{\pi} = \boldsymbol{\pi}_0$ with no parameter estimation), a suitably normalized X^2 statistic ($\lambda = 1$) has a limiting standard normal distribution. For $\lambda > -1$, the asymptotic normalizing mean and variance of the power-divergence statistic $2nI^{\lambda}(\mathbf{X}/n : \boldsymbol{\pi}_0)$ can be calculated to be

$$\mu_\lambda = 2nI^{\lambda}(\boldsymbol{\pi}_1 : \boldsymbol{\pi}_0)$$

and

$$\sigma_\lambda^2 = \frac{4n}{\lambda^2}\left[\sum_{i=1}^{k}\left(\frac{\pi_{1i}}{\pi_{0i}}\right)^{2\lambda}\pi_{1i} - \left[\sum_{i=1}^{k}\left(\frac{\pi_{1i}}{\pi_{0i}}\right)^{\lambda}\pi_{1i}\right]^2\right],$$

provided $\lambda \neq 0$ (Section A9 of the Appendix). Here $\boldsymbol{\pi}_0$ and $\boldsymbol{\pi}_1$ are the cell probabilities under the simple null and (fixed) alternative model, respectively (a similar result can be derived in the case $\lambda = 0$, but not for $\lambda \leq -1$ since the moments do not exist).

Approximating the Small-Sample Power

Both the noncentral chi-squared and the normal asymptotic results can be used to approximate the exact power of the test for a given sample size. Broffitt and Randles (1977) perform Monte Carlo simulations that indicate for large values of the exact power, the normal approximation is better, otherwise the chi-squared approximation should be used. Their simulations are of course only for Pearson's X^2 ($\lambda = 1$), and it would be interesting to see how the statistic $2nI^{2/3}(\mathbf{X}/n : \boldsymbol{\pi}_0)$, proposed in Section 1.2 behaves under similar simulations. In this case the asymptotic mean and variance that would be used are

$$\mu_{2/3} = 2nI^{2/3}(\boldsymbol{\pi}_1 : \boldsymbol{\pi}_0)$$

and

$$\sigma_{2/3}^2 = 9n\left[\sum_{i=1}^{k}\left(\frac{\pi_{1i}}{\pi_{0i}}\right)^{4/3}\pi_{1i} - \left[\sum_{i=1}^{k}\left(\frac{\pi_{1i}}{\pi_{0i}}\right)^{2/3}\pi_{1i}\right]^2\right].$$

(Bishop et al., 1975, pp. 518–519, provide a brief discussion of asymptotic distributions for general distance measures assuming nonlocal alternative models and estimated parameters.)

Drost et al. (1987) derive two new approximations to the power function of the power-divergence statistic which are based on Taylor-series expansions valid for nonlocal alternatives. Broffitt and Randles' (1977) approximation is a special case (after moment correction); for small samples, neither the noncentral chi-squared approximation nor the approximation of Broffitt and Randles performs as well as these new approximations. Drost et al. analyze the form of their new approximations, and conclude that large values of λ are most efficient in detecting alternatives for which the more important contributions have large ratios π_{1i}/π_{0i}. Small values of λ are preferable for detecting near-zero ratios of π_{1i}/π_{0i}. These observations concur with the results of Section 5.4 where we obtain exact power calculations for some specific alternative models and provide a more detailed analysis of the relative efficiencies of the power-divergence family members.

4.3. Significance Levels and Efficiency under Sparseness Assumptions

The development of large-sample distributions in the previous two sections assumes implicitly that as the sample size n increases, the number of multinomial cells k remains fixed. Consequently, for approximation purposes, these asymptotics assume k small relative to n. Since $\boldsymbol{\pi}_0$ is fixed, it follows that under the model H_0: $\boldsymbol{\pi} = \boldsymbol{\pi}_0$ in (4.2), each element of the expected frequency vector $n\boldsymbol{\pi}_0$ will be large.

While the main emphasis of the literature has been to use this asymptotic machinery, it is noted by Holst (1972) that "it is rather unnatural to keep [k] fixed when $n \to \infty$ [for the traditional goodness-of-fit problem of testing whether a sample has come from a given population]; instead we should have that [k] $\to \infty$ when $n \to \infty$." In the case of contingency table analyses, Fienberg (1980, pp. 174–175) states "The fact remains ... that with the extensive questionnaires of modern-day sample surveys, and the detailed and painstaking inventory of variables measured by biological and social scientists, the statistician is often faced with large sparse arrays, full of 0's and 1's, in need of careful analysis."

Under such a scheme where k increases without limit, it must be remembered that the dimension and structure of the probability space is changing with k. Consequently the expected cell frequencies are no longer assured of becoming large with n, as required for the application of the classical (fixed-cells) asymptotic theory where the cell probabilities are fixed for all n. An indication of the need to investigate this situation is given by Hoeffding (1965, pp. 371–372). He comments that his results indicating the global equivalence or superiority of the loglikelihood ratio test G^2 ($\lambda = 0$) over Pearson's X^2 ($\lambda = 1$) "are subject to the limitation that k is fixed or does not increase rapidly with n. Otherwise the relation between the two tests may be reversed." An example follows in which it is assumed that the ratio n/k is "moderate," and Hoeffding points out that X^2 is superior to G^2 for alternatives local to the null model.

The necessary asymptotic theory has been developed under slightly different sparseness assumptions by Morris (1966, 1975) and Holst (1972) (more recently Dale, 1986, has extended the theory for product-multinomial sampling; Koehler, 1986, considers models for sparse contingency tables requiring parameter estimation, which we discuss in Section 8.1). Under restrictions on the rate at which $k \to \infty$ (the ratio n/k must remain finite), it has been shown that X^2 and G^2 have different asymptotic normal distributions. The asymptotic normality we might expect intuitively, since for fixed k the asymptotic distribution of both statistics is chi-squared with degrees of freedom proportional to k. Increasing the degrees of freedom of a chi-squared random variable results in it approaching (in distribution) a normal random variable. The surprising feature is that the asymptotic mean and asymptotic standardized variance differ for X^2 and G^2. We now describe these results in more detail, and extend them to the other members of the power-divergence family.

The Equiprobable Model

First we consider the case where the equiprobable model is hypothesized. As before, we assume that $\mathbf{X}$ is a multinomial probability vector, but because we let $k \to \infty$, we need to explicitly notate the cell probabilities and the sample size n as functions of k. Hence we write: $\mathbf{X}_k = (X_{1k}, X_{2k}, \ldots, X_{kk})$ is multinomial $\text{Mult}_k(n_k, \boldsymbol{\pi}_k)$, and hypothesize the null model

$$H_0\colon \boldsymbol{\pi}_k = \mathbf{1}/k \tag{4.15}$$

where $\mathbf{1} = (1, 1, \ldots, 1)$ is a vector of length k. This model is an important special case of (4.2) as discussed in the Historical Perspective and by Read (1984b).

Using the sparseness assumptions of Holst (1972), we state the following result (proved in Section A10 of the Appendix). Suppose $n_k \to \infty$ as $k \to \infty$ so that $n_k/k \to a$ $(0 < a < \infty)$. Assume hypothesis (4.15) holds and $\lambda > -1$; then for any $c \geq 0$

$$\begin{aligned} &Pr[(2n_k I^\lambda(\mathbf{X}_k/n_k : \mathbf{1}/k) - \mu_k^{(\lambda)})/\sigma_k^{(\lambda)} \geq c] \\ &\quad \to Pr[N(0,1) \geq c], \qquad \text{as } k \to \infty, \end{aligned} \tag{4.16}$$

where $N(0,1)$ represents a standard normal random variable and

$$\mu_k^{(\lambda)} = \begin{cases} [2n_k/(\lambda(\lambda+1))]E\{(Y_k/m_k)^{\lambda+1} - 1\}; & \lambda > -1, \lambda \neq 0 \\ [2n_k]E\{(Y_k/m_k)\log(Y_k/m_k)\}; & \lambda = 0, \end{cases}$$

$$[\sigma_k^{(\lambda)}]^2 = \begin{cases} [2m_k/(\lambda(\lambda+1))]^2 k[\operatorname{var}\{(Y_k/m_k)^{\lambda+1}\} \\ \quad - m_k \operatorname{cov}^2\{Y_k/m_k, (Y_k/m_k)^{\lambda+1}\}]; & \lambda > -1, \lambda \neq 0 \\ [2m_k]^2 k[\operatorname{var}\{(Y_k/m_k)\log(Y_k/m_k)\} \\ \quad - m_k \operatorname{cov}^2\{Y_k/m_k, (Y_k/m_k)\log(Y_k/m_k)\}]; & \lambda = 0; \end{cases}$$

and Y_k is a Poisson random variable with mean $m_k = n_k/k$.

This result indicates that under such sparseness assumptions, the members of the power-divergence family are no longer asymptotically equivalent (as they are in Section 4.1). In the case of a zero observed cell frequency, $2n_k I^\lambda(\mathbf{x}_k/n_k : \boldsymbol{\pi}_k)$ is undefined for $\lambda \leq -1$, since it requires taking positive powers of $n_k \pi_{ik}/x_{ik}$ where $x_{ik} = 0$. Similarly $\mu_k^{(\lambda)}$ and $[\sigma_k^{(\lambda)}]^2$ in (4.16) are not defined for $\lambda \leq -1$, because Y_k has a positive probability of equaling 0.

Efficiency for Local Alternatives to the Equiprobable Model

How efficient are the various members of the power-divergence family under these new asymptotics? Recall from Section 4.2, the family members are asymptotically equally efficient for testing against local alternative models that converge to the null. Here this equivalence no longer holds. In particular, for testing the model

$$H_0: \boldsymbol{\pi}_k = \mathbf{1}/k$$

versus (4.17)

$$H_{1,k}: \boldsymbol{\pi}_k = \mathbf{1}/k + \boldsymbol{\delta}/n_k^{1/4},$$

where $\boldsymbol{\delta} = (\delta_1, \delta_2, \ldots, \delta_k)$ and $\sum_{i=1}^{k} \delta_i = 0$, the power-divergence statistic has a normal distribution also under $H_{1,k}$ (Section A10 of the Appendix) and the Pitman asymptotic efficiency of $2n_k I^\lambda(\mathbf{X}_k/n_k : \mathbf{1}/k)$ is proportional to $\sqrt{(a/2)}\rho_\lambda \equiv \sqrt{(a/2)}\,\mathrm{sgn}(\lambda)\,\mathrm{corr}\{Y^{\lambda+1} - a^{-1}\,\mathrm{cov}(Y^{\lambda+1}, Y)Y,\ \ Y^2 - (2a+1)Y\}$ where Y is Poisson with mean a. Holst (1972) shows that $\rho_1 = 1$, and since $\rho_\lambda \le 1$, Pearson's X^2 test ($\lambda = 1$) will be maximally efficient among the power-divergence family for testing (4.17). The efficiency losses resulting from using λ other than $\lambda = 1$, are illustrated in Table 4.5. For $\lambda > 3$, the efficiency drops off rapidly, but for $-1 < \lambda \le 3$ the efficiencies are close to optimal. For large values of a, that is, for n_k large relative to k, the changes in the tabled values with λ are small. This coincides with the results of Sections 4.1 and 4.2, where we proved the statistics to be equivalent under these conditions.

For the equiprobable null hypothesis, Koehler and Larntz (1980) perform Monte Carlo power comparisons of Pearson's X^2 ($\lambda = 1$) and the loglikelihood ratio G^2 ($\lambda = 0$) tests when n_k/k is moderate (i.e., in the range 1/4 to 5). They conclude, in agreement with the work of this section, that X^2 is slightly more powerful for local alternatives. However, they propose further that for more distant alternatives with one or two cells having very small probabilities

Table 4.5. Values of $\sqrt{(a/2)}\rho_\lambda$ for various λ and a

	a									
λ	0.0	0.1	0.5	1.0	1.5	2.0	3.0	10.0	20.0	50.0
−2/3	0.00	0.22	0.47	0.62	0.72	0.78	0.89	2.07	3.08	4.95
−1/2	0.00	0.22	0.47	0.63	0.74	0.83	0.98	2.12	3.10	4.96
−1/3	0.00	0.22	0.48	0.65	0.77	0.87	1.05	2.15	3.11	4.97
0	0.00	0.22	0.49	0.67	0.81	0.93	1.15	2.19	3.13	4.98
1/3	0.00	0.22	0.49	0.69	0.84	0.97	1.20	2.22	3.15	4.99
1/2	0.00	0.22	0.50	0.70	0.85	0.99	1.21	2.23	3.16	5.00
2/3	0.00	0.22	0.50	0.70	0.86	0.99	1.22	2.23	3.16	5.00
1	0.00	0.22	0.50	0.71	0.87	1.00	1.22	2.24	3.16	5.00
2	0.00	0.22	0.48	0.68	0.83	0.96	1.19	2.21	3.14	4.98
3	0.00	0.21	0.43	0.61	0.76	0.89	1.10	2.13	3.07	4.94
4	0.00	0.18	0.37	0.53	0.67	0.79	1.00	2.01	2.97	4.86
5	0.00	0.15	0.30	0.44	0.57	0.67	0.87	1.87	2.85	4.76

Source: Cressie and Read (1984).

and the rest reasonably equal, G^2 is more powerful. But as the number of near-zero alternative probabilities increases, X^2 dominates again. This behavior reinforces our general recommendation of Section 4.5 to use $\lambda = 2/3$ as a good compromise. This important point is discussed further in Section 6.3.

The Model for Unequal Cell Probabilities

When the cell probabilities in the null model are unequal, the results of Morris (1975) can be used to illustrate a more general result than (4.16). Here we consider the null model

$$H_0: \boldsymbol{\pi}_k = \boldsymbol{\pi}_{0k} \tag{4.18}$$

(i.e., $\pi_{ik} = \pi_{0ik}$, for $i = 1, \ldots, k$). We assume that $n_k \to \infty$ as $k \to \infty$ and we place several restrictions on $\boldsymbol{\pi}_k$ and $\boldsymbol{\pi}_{0k}$, namely: $\pi_{0ik} > 0$ for all $i = 1, \ldots, k$, and $k > 0$; $\sum_{i=1}^{k} \pi_{0ik} = 1$; $\max_{1 \le i \le k} \pi_{ik} = o(1)$ (i.e., no single cell dominates the multinomial probabilities); and $n_k \pi_{ik} \ge \varepsilon$, for some $\varepsilon > 0$, and for all $i = 1, \ldots, k$, and $k > 0$. (i.e., all expected cell frequencies are nonzero). Furthermore, we consider only the integer values of λ, $\lambda = 0, 1, 2, \ldots$. Then writing

$$2n_k I^\lambda(\mathbf{X}_k/n_k : \boldsymbol{\pi}_{0k}) = \frac{2n_k}{\lambda(\lambda+1)} \sum_{i=1}^{k} \pi_{0ik}\left[\left(\frac{X_{ik}}{n_k \pi_{0ik}}\right)^{\lambda+1} - 1\right]$$

it follows that for $\lambda = 0, 1, 2, 3, \ldots$

$$Pr[(2n_k I^\lambda(\mathbf{X}_k/n_k : \boldsymbol{\pi}_{0k}) - \mu_k^{(\lambda)})/\sigma_k^{(\lambda)} \ge c] \to Pr[N(0,1) \ge c], \qquad \text{as } k \to \infty, \tag{4.19}$$

for any $c \ge 0$ provided

$$\max_{1 \le i \le k} [\sigma_{ik}^{(\lambda)}]^2/[\sigma_k^{(\lambda)}]^2 = o(1), \qquad \text{as } k \to \infty, \tag{4.20}$$

where

$$\mu_k^{(\lambda)} = \begin{cases} \dfrac{2n_k}{\lambda(\lambda+1)} \displaystyle\sum_{i=1}^{k} \pi_{0ik} E\left[\left(\frac{Y_{ik}}{n_k \pi_{0ik}}\right)^{\lambda+1} - 1\right]; & \lambda = 1, 2, \ldots \\ 2n_k \displaystyle\sum_{i=1}^{k} \pi_{0ik} E\left[\frac{Y_{ik}}{n_k \pi_{0ik}} \log\left(\frac{Y_{ik}}{n_k \pi_{0ik}}\right)\right]; & \lambda = 0, \end{cases}$$

$$[\sigma_{ik}^{(\lambda)}]^2 = \begin{cases} \left(\dfrac{2n_k}{\lambda(\lambda+1)}\right)^2 (\pi_{0ik})^2 \operatorname{var}\left[\left(\dfrac{Y_{ik}}{n_k \pi_{0ik}}\right)^{\lambda+1}\right] \\ \quad + [\gamma_k^{(\lambda)}]^2 n_k \pi_{ik} - 2\gamma_k^{(\lambda)} n_k \gamma_{ik}^{(\lambda)}; \quad \lambda = 1, 2, \ldots \\ (2n_k)^2 (\pi_{0ik})^2 \operatorname{var}\left[\dfrac{Y_{ik}}{n_k \pi_{0ik}} \log\left(\dfrac{Y_{ik}}{n_k \pi_{0ik}}\right)\right] \\ \quad + [\gamma_k^{(\lambda)}]^2 n_k \pi_{ik} - 2\gamma_k^{(\lambda)} n_k \gamma_{ik}^{(\lambda)}; \quad \lambda = 0, \end{cases}$$

$$[\sigma_k^{(\lambda)}]^2 = \sum_{i=1}^{k} [\sigma_{ik}^{(\lambda)}]^2;$$

$$\gamma_{ik}^{(\lambda)} = \begin{cases} \dfrac{2n_k(\pi_{0ik})^2}{\lambda(\lambda+1)} \operatorname{cov}\left[\dfrac{Y_{ik}}{n_k\pi_{0ik}}, \left(\dfrac{Y_{ik}}{n_k\pi_{0ik}}\right)^{\lambda+1}\right]; & \lambda = 1, 2, \ldots \\ 2n_k(\pi_{0ik})^2 \operatorname{cov}\left[\dfrac{Y_{ik}}{n_k\pi_{0ik}}, \dfrac{Y_{ik}}{n_k\pi_{0ik}} \log\left(\dfrac{Y_{ik}}{n_k\pi_{0ik}}\right)\right]; & \lambda = 0, \end{cases}$$

$$\gamma_k^{(\lambda)} = \sum_{i=1}^{k} \gamma_{ik}^{(\lambda)};$$

and the Y_{ik}s are independent Poisson random variables with means $n_k\pi_{ik}$; $i = 1, \ldots, k$.

The proof of (4.19) follows from corollary 4.1 and theorem 5.2 of Morris (1975). So far (4.19) is proved only when λ is a nonnegative integer, due to the restriction to polynomials in $\mathbf{X}_k$ of Morris' corollary 4.1. However the continuity of $2n_k I^\lambda(\mathbf{X}_k/n_k : \boldsymbol{\pi}_{0k})$ in λ should ensure that (4.19) will hold for all real $\lambda > -1$. This result has yet to be proved rigorously.

When hypothesis (4.18) is true (i.e., $\pi_{ik} = \pi_{0ik}$; $i = 1, \ldots, k$), condition (4.20) is automatically satisfied. Furthermore when $\pi_{0ik} = 1/k$; $i = 1, \ldots, k$ then $\mu_k^{(\lambda)}$ and $[\sigma_k^{(\lambda)}]^2$ of (4.19) reduce to $\mu_k^{(\lambda)}$ and $[\sigma_k^{(\lambda)}]^2$ of (4.16).

The distributional result (4.19) provides a basis for efficiency comparisons between the members of the power-divergence family similar to those discussed already for the equiprobable hypothesis, where it was shown that Pearson's X^2 is optimal. However, for the general hypothesis (4.18), Ivchenko and Medvedev (1978) illustrate through some specific examples that no such uniform optimality exists. Furthermore Haberman (1986) points out that, for sparse contingency tables, X^2 can exhibit serious bias when testing hypotheses containing unequal cell probabilities (Section 3 of the Historical Perspective). Further comments on optimal tests are contained in Section 8.1 together with a discussion of different methods of parameter estimation.

Comparing Models with Constant Difference in Degrees of Freedom

Haberman (1977) examines the distribution of X^2 ($\lambda = 1$) and G^2 ($\lambda = 0$) defined for comparing two loglinear models in large sparse contingency tables, where one model is a special case of the other (i.e., hierarchical models). Under some specific regularity conditions both X^2 and G^2 converge to chi-squared random variables with d degrees of freedom, where d is the difference in the number of parameters for the two models (Section 8.1). Therefore if our primary interest in analyzing a large sparse table is to focus on assessing the importance of a fixed finite subset of parameters, then we should try to formulate two models that differ only by those parameters and test their significance using the chi-squared distribution (Agresti and Yang, 1987). We expect Haberman's result will hold for all the members of the power-divergence family (Section 8.1).

4.4. A Summary Comparison of the Power-Divergence Family Members

The results of the previous sections illustrate that the optimal choice of λ for the power-divergence test statistic depends on (a) how the model changes with sample size n, and (b) the structure of the null model under test together with any alternative models against which we wish to discriminate.

In Section 4.1 it was shown that under the classical (fixed-cells) assumptions all members of the power-divergence family are asymptotically equivalent as $n \to \infty$, provided the null model is true. Their common asymptotic distribution is the chi-squared distribution with degrees of freedom depending on k and the number of BAN-estimated parameters. When the null model is not true, but is replaced by a local alternative (in the sense that it converges to the null model at the rate $n^{-1/2}$, as $n \to \infty$; see Section 4.2), this equivalence still holds although the limiting distribution is now noncentral chi-squared. This result indicates that all power-divergence family members are asymptotically equally efficient for detecting such local alternative models.

For nonlocal alternatives that are a fixed distance away from the null model (as $n \to \infty$), the members of the power-divergence family are no longer equivalent. Their limiting distribution changes from a noncentral chi-squared to a standard normal after appropriate standardization, where the standardizing mean and variance depend on λ and the true cell probabilities. Consequently, the asymptotic efficiency for detecting such alternatives is different for each λ in the power-divergence family, and depends on the relative magnitudes of the ratios of the null and alternative cell probabilities for each cell. No general rule emerges as to which λ is dominant for which alternative models, however some specific examples of differences are discussed in Sections 5.4 and 6.3.

A different way of defining efficiency for nonlocal alternatives is to examine the rate at which the attained significance level tends to 0 when the alternative model is true. This leads to the concept of Bahadur efficiency (Section 4.2), which indicates the loglikelihood ratio test G^2 ($\lambda = 0$) is optimal in this sense.

In Section 4.3 the number of cells k was made to grow at a similar rate to the sample size n. In this case the power-divergence family members are no longer asymptotically equivalent under the null hypothesis, but obtain a normal distribution with mean and variance dependent on both λ and the hypothesized cell probabilities. Efficiency comparisons for the equiprobable null model against local alternatives indicate that Pearson's X^2 ($\lambda = 1$) is optimal, however the loss in efficiency is not great until λ is two units away from 1. For alternatives further away from the null model, Pearson's X^2 does not always dominate. For null models in which the cell probabilities are not all equal, no uniform optimality exists for a specific λ; each case must be treated individually (Section 8.1).

4.5. Which Test Statistic?

Clearly there are some conflicting recommendations regarding which value of λ results in the optimal test statistic. This is due to the wide range of conditions under which goodness-of-fit tests are used and the varied criteria for optimality which have been considered in the literature. In almost all cases a reasonable choice of λ will lie in the range $\lambda \in (-1, 2]$. This conclusion is based on not only the results of this chapter, but also on those of Chapter 5, where the calculation of significance levels for small samples is examined.

If the null and alternative models we wish to consider are very specific, a detailed analysis of efficiency from the results presented so far will indicate the optimal λ. For example: $\lambda = 1$ for testing the equiprobable hypothesis against certain local alternatives in large sparse tables; $\lambda = 0$ for testing against certain nonlocal alternatives with some near-zero probabilities. However, if there is little or no knowledge of possible alternative models, then a good compromise, reinforced by the results of Chapters 5 and 6, is to choose $\lambda = 2/3$; that is

$$2nI^{2/3}(\mathbf{x}/n : \hat{\boldsymbol{\pi}}) = \frac{9}{5} \sum_{i=1}^{k} x_i \left[\left(\frac{x_i}{n\hat{\pi}_i} \right)^{2/3} - 1 \right].$$

The discussion in Chapter 5 proposes that provided $\min_{1 \leq i \leq k} n\hat{\pi}_i \geq 1$ and $n \geq 10$, we can obtain a good approximation to the critical value c_α by using $\chi^2_{k-s-1}(\alpha)$ from the chi-squared tables (where s is the number of BAN-estimated parameters). If the expected frequencies are all nearly equal, then preliminary results indicate the chi-squared approximation may still be accurate for expected frequencies as low as 1/4 (provided $n^2/k \geq 10$, $n \geq 10$ and $k \geq 3$). If some of the expected frequencies are very small while others are greater than 1, there are no firm recommendations regarding the best approximation to use for the critical value. In some of these cases the normal approximation of Section 4.3 will be appropriate, however the convergence of this normal approximation is slow. This is a topic for further research (Section 8.1).

CHAPTER 5

Improving the Accuracy of Tests with Small Sample Size

The distributional properties of the power-divergence family discussed in Chapters 3 and 4 rely on large sample sizes for their validity (i.e., they are asymptotic results). So far, we have not discussed the relevance of these properties when the sample size is small.

This chapter explores the accuracy of applying asymptotic results in cases where the sample size cannot be assumed large. How far is the asymptotic significance level from the exact significance level for the power-divergence statistic? Are there some more accurate approximations we can use for small samples? When should we regard a sample as small? Is the small-sample relative efficiency of the power-divergence family members similar to the large-sample results of Sections 4.2 and 4.3?

Sections 5.1 through 5.4 provide answers to these questions, while Section 5.5 summarizes the results and provides specific recommendations as to which test statistics are best for small samples.

5.1. Improved Accuracy through More Accurate Moments

In order to assess the small-sample accuracy of asymptotic significance levels, we shall begin by comparing the asymptotic moments of the test statistic with small-sample expressions for these moments. In Section 5.3 we shall see that the closeness of the exact mean and variance to the asymptotic mean and variance of the power-divergence statistic reflects the closeness of the exact significance level to the asymptotic significance level for this test statistic.

For simplicity, first we deal with the null model

$$H_0\colon \boldsymbol{\pi} = \boldsymbol{\pi}_0, \tag{5.1}$$

where $\boldsymbol{\pi}_0$ is a completely specified probability vector, and the number of cells k is fixed. In Section A11 of the Appendix the first three moments of $2nI^\lambda(\mathbf{X}/n : \boldsymbol{\pi}_0)$ for $\lambda > -1$, defined from (4.1), are expanded to include terms of order n^{-1} (for $\lambda \leq -1$ the moments do not exist; see, e.g., Bishop et al., 1975, p. 488). From these expansions, the mean and variance for $\lambda > -1$ can be computed to be:

$$\begin{aligned} E[2nI^\lambda(\mathbf{X}/n : \boldsymbol{\pi}_0)] = [k-1] &+ n^{-1}[(\lambda - 1)(2 - 3k + t)/3 \\ &+ (\lambda - 1)(\lambda - 2)(1 - 2k + t)/4] + o(n^{-1}) \end{aligned} \tag{5.2}$$

and

$$\begin{aligned} \operatorname{var}[2nI^\lambda(\mathbf{X}/n : \boldsymbol{\pi}_0)] = [2k-2] &+ n^{-1}[(2 - 2k - k^2 + t) \\ &+ (\lambda - 1)(8 - 12k - 2k^2 + 6t) \\ &+ (\lambda - 1)^2(4 - 6k - 3k^2 + 5t)/3 \\ &+ (\lambda - 1)(\lambda - 2)(2 - 4k + 2t)] + o(n^{-1}), \end{aligned} \tag{5.3}$$

where $t = \sum_{i=1}^{k} \pi_{0i}^{-1}$ (special cases of these formulas are given in Haldane, 1937 or Johnson and Kotz, 1969, p. 286 for $\lambda = 1$, and in Smith et al., 1981, for $\lambda = 0$).

Choosing λ to Minimize the Correction Terms

The first terms on the right-hand sides of (5.2) and (5.3) are the mean and variance, respectively, of a chi-squared random variable with $k - 1$ degrees of freedom. The second terms are the correction terms of order n^{-1}; they involve the family parameter λ, the number of cells k (assumed fixed), and the sum of the reciprocal null probabilities t. If we denote the correction term for the mean by $f_m(\lambda, k, t)$ and the correction term for the variance by $f_v(\lambda, k, t)$,

$$\begin{aligned} E[2nI^\lambda(\mathbf{X}/n : \boldsymbol{\pi}_0)] &= k - 1 + n^{-1} f_m(\lambda, k, t) + o(n^{-1}) \\ \operatorname{var}[2nI^\lambda(\mathbf{X}/n : \boldsymbol{\pi}_0)] &= 2k - 2 + n^{-1} f_v(\lambda, k, t) + o(n^{-1}), \end{aligned} \tag{5.4}$$

then it is clear that f_m and f_v control the speed at which the mean and variance of the power-divergence statistic converge to the mean and variance of a chi-squared random variable with $k - 1$ degrees of freedom. Consequently we are interested in finding the values of $\lambda > -1$ for which f_m and f_v are close to 0. For fixed k and t, (5.2) and (5.3) show that $f_m(\lambda, k, t)$ and $f_v(\lambda, k, t)$ are quadratics in λ, so we can solve directly for the two solutions to each of the equations $f_m(\lambda, k, t) = 0$ and $f_v(\lambda, k, t) = 0$.

For the special case of the equiprobable hypothesis in (5.1),

$$H_0\colon \boldsymbol{\pi} = \mathbf{1}/k, \tag{5.5}$$

$t = \sum_{i=1}^{k} \pi_{0i}^{-1} = k^2$. Hence for $k \geq 2$

$$f_m(\lambda, k, k^2) = 0, \qquad \text{when } \lambda = 1 \text{ or } \lambda = 2 - \frac{4(k-2)}{3(k-1)}$$

and

$$f_v(\lambda, k, k^2) = 0, \qquad \text{when } \lambda = \frac{5k - 1 \pm [3(3k^2 - 2k + 7)]^{1/2}}{2(4k-5)}.$$

Table 5.1 gives the two solutions to each of $f_m(\lambda, k, k^2) = 0$ and $f_v(\lambda, k, k^2) = 0$ for increasing values of k, from which we see that these solutions are fairly insensitive to the choice of k. For $k > 50$ the solutions are essentially constant. Clearly $\lambda = 1$ (Pearson's X^2) minimizes the correction term for the mean over all k. However $\lambda = 1$ minimizes the correction term for the variance only for large values of k.

For arbitrary completely specified hypotheses of the form (5.1), we no longer have $t = \sum_{i=1}^{k} \pi_{0i}^{-1} = k^2$. In general, $t \geq k^2$ since k^2 is the minimum value of t under the constraints $\pi_{0i} > 0$; $i = 1, \ldots, k$ and $\sum_{i=1}^{k} \pi_{0i} = 1$. How do large values of t (i.e., values of t of larger order than k^2) affect Table 5.1? We can answer this question by looking back at (5.2) and (5.3) and treating the quadratics in k as negligible compared to t. Then

$$f_m(\lambda, k, t) = 0 \qquad \text{when } \lambda = 1 \text{ or } \lambda = 2/3$$

and

$$f_v(\lambda, k, t) = 0 \qquad \text{when } \lambda = 0.30 \text{ or } \lambda = 0.61.$$

Summarizing, we see that for the equiprobable hypothesis (5.5), where $t = k^2$, Pearson's X^2 ($\lambda = 1$) tends to produce the smallest correction terms for $k \geq 20$; but in the cases where t dominates k^2, choosing $\lambda \in [0.61, 0.67]$ results in the smallest mean and variance correction terms. The importance of the parameter value $\lambda = 2/3$ emerges also when considering the third moment as described in Section A11 of the Appendix.

Table 5.1. Entries Show the Solutions in λ to $f_m = 0$ and $f_v = 0$ from (5.4) for the Equiprobable Hypothesis (5.5) as k Increases

Correction term	k: 2	3	4	5	10	20	50	100	$\cdots$	∞
$f_m(\lambda, k, k^2)$	1.00	1.00	1.00	1.00	0.81	0.74	0.69	0.68	$\cdots$	2/3
$f_m(\lambda, k, k^2)$	2.00	1.33	1.11	1.00	1.00	1.00	1.00	1.00	$\cdots$	1
$f_v(\lambda, k, k^2)$	0.38	0.35	0.32	0.31	0.28	0.27	0.26	0.25	$\cdots$	1/4
$f_v(\lambda, k, k^2)$	2.62	1.65	1.40	1.29	1.12	1.05	1.02	1.01	$\cdots$	1

The Moment-Corrected Statistic

For a given critical value c, define the *distribution tail function* of the chi-squared distribution to be

$$T_\chi(c) = Pr(\chi^2_{k-1} \geq c), \tag{5.6}$$

where χ^2_{k-1} is a chi-squared random variable with $k - 1$ degrees of freedom; then $T_\chi(c)$ is the asymptotic significance level associated with the critical value c. An ad hoc method of improving the small-sample accuracy of this approximation to the significance level is to define a moment-corrected distribution tail function, based on the moment-corrected statistic

$$\{2nI^\lambda(\mathbf{X}/n : \boldsymbol{\pi}_0) - \mu_\lambda\}/\sigma_\lambda; \qquad -\infty < \lambda < \infty,$$

with

$$\mu_\lambda = (k-1)(1-\sigma_\lambda) + f_m(\lambda, k, t)/n$$
$$\sigma^2_\lambda = 1 + f_v(\lambda, k, t)/(2(k-1)n),$$

and

$$f_m(\lambda, k, t) = (\lambda - 1)(2 - 3k + t)/3 + (\lambda - 1)(\lambda - 2)(1 - 2k + t)/4$$
$$f_v(\lambda, k, t) = 2 - 2k - k^2 + t + (\lambda - 1)(8 - 12k - 2k^2 + 6t)$$
$$+ (\lambda - 1)^2(4 - 6k - 3k^2 + 5t)/3 + (\lambda - 1)(\lambda - 2)(2 - 4k + 2t);$$

where $t = \sum_{i=1}^k \pi_{0i}^{-1}$. This corrected statistic will have mean and variance (for $\lambda > -1$) matching the chi-squared mean and variance, $k - 1$ and $2(k - 1)$, respectively, to $o(n^{-1})$. While the mean and variance of $2nI^\lambda(\mathbf{X}/n : \boldsymbol{\pi}_0)$ do not exist for $\lambda \leq -1$, the corrected statistic is still well defined, and we define the *moment-corrected* distribution tail function as

$$T_C(c) = T_\chi((c - \mu_\lambda)/\sigma_\lambda), \tag{5.7}$$

where T_χ comes from (5.6). For $\lambda > -1$, T_C should provide a more accurate approximation to the small-sample significance level of the test based on the power-divergence statistic. In Section 5.3 we illustrate numerically that T_C does indeed result in a substantial improvement in accuracy for values of λ outside the interval $[1/3, 3/2]$, for both $\lambda > -1$ and $\lambda \leq -1$.

Application of the Moment-Correction Terms

The second-order mean and variance approximations derived in this section (under the classical (fixed-cells) assumptions of Section 4.1) serve two purposes. One is to obtain more accurate moment formulas that are no longer equivalent for all values of the family parameter $\lambda > -1$. These correction terms provide an important way to distinguish between the asymptotically equivalent power-divergence family members, and tie in with the results of Section 5.3 to illustrate that family members in the range $\lambda \in [1/3, 3/2]$ have distribution tail functions that converge most rapidly to the chi-squared distribution tail function (5.6). The second purpose is to provide the corrected distribution tail function (5.7) which reflects the exact distribution tail function of the power-divergence statistic more accurately when $\lambda \notin [1/3, 3/2]$ (Section 5.3); values of λ for which the moment-correction terms are not negligible. (For second-order moment approximations under the sparseness assumptions of Section 4.3, see Section 8.1.)

In this section we have assumed that the null hypothesis is completely specified. The effect of parameter estimation on the small-sample accuracy of asymptotic significance levels provides some interesting open questions for subsequent comparisons of the power-divergence family members. These are discussed briefly at the end of Section 5.3 and in Section 8.1.

5.2. A Second-Order Correction Term Applied Directly to the Asymptotic Distribution

We shall now consider a more direct and mathematically rigorous approach to assessing the adequacy of the chi-squared approximation to the distribution function of the power-divergence statistic under the asymptotics of Section 4.1. Assuming the simple null model (5.1), define the *second-order-corrected* distribution tail function for the power-divergence statistic to be

$$T_S(c) = T_\chi(c) + J_1^\lambda(c) + J_2^\lambda(c), \tag{5.8}$$

where T_χ is the chi-squared distribution tail function with $k - 1$ degrees of freedom from (5.6), and J_1^λ and J_2^λ are second-order correction terms (derived in Section A12 of the Appendix).

The term J_1^λ is the standard Edgeworth expansion term used for continuous distribution functions and is given by

$$\begin{aligned} J_1^\lambda(c) = (24n)^{-1}\{&Pr(\chi_{k-1}^2 < c)[2(1-t)] \\ &+ Pr(\chi_{k+1}^2 < c)[-3(2k + k^2 - 3t) - (\lambda-1)6(k^2 - t) \\ &+ (\lambda-1)^2(4-6k-3k^2+5t) - (\lambda-1)(\lambda-2)3(1-2k+t)] \\ &+ Pr(\chi_{k+3}^2 < c)[-6(1-2k-k^2+2t) - (\lambda-1)4(2-3k-3k^2+4t) \\ &- (\lambda-1)^2 2(4-6k-3k^2+5t) + (\lambda-1)(\lambda-2)3(1-2k+t)] \\ &+ Pr(\chi_{k+5}^2 < c)[\lambda^2(4-6k-3k^2+5t)]\}, \end{aligned} \tag{5.9}$$

where χ_v^2 represents a chi-squared random variable with v degrees of freedom, and $t = \sum_{i=1}^k \pi_{0i}^{-1}$. Due to the discreteness of the distribution function for the power-divergence statistic, the Edgeworth correction term J_1^λ is not sufficient as a full second-order correction term. The further term J_2^λ accounts for the error due to discontinuity, and is estimated by

$$J_2^\lambda(c) = \frac{\left(\begin{bmatrix}\#\text{ lattice points}\\ \text{in } B_\lambda(c)\end{bmatrix} - \begin{bmatrix}\text{volume of}\\ B_\lambda(c)\end{bmatrix}\right) n^{(k-1)/2}}{\exp(c/2)\left[(2\pi n)^{k-1}\prod_{i=1}^{k}\pi_{0i}\right]^{1/2}}, \tag{5.10}$$

where $B_\lambda(c)$ is the set of all vectors $(w_1, w_2, \ldots, w_{k-1})$ which satisfy $w_i = \sqrt{n}(x_i/n - \pi_{0i})$; $x_i = 0, 1, 2, \ldots$ such that $\sum_{i=1}^{k} x_i = n$ and $2nI^\lambda(\mathbf{x}/n : \boldsymbol{\pi}_0) < c$.

Compared to the correction terms for the mean and variance in (5.2) and (5.3), the correction terms (5.9) and (5.10) provide a more direct assessment of the adequacy of approximating the distribution tail function of the power-divergence statistic by the chi-squared distribution tail function. However the effort required to calculate (5.9) and (5.10) is substantial in comparison to calculating (5.2) and (5.3). Fortunately, as we shall see in the next section, the moment-corrected distribution tail function (5.7) is as accurate as the more complicated formula (5.8) in almost all the cases we consider.

A similar second-order correction could now be pursued under the asymptotics of Section 4.3, where both n and k become large (Section 8.1). However from the practical point of view of accuracy and computability, the critical values obtained in the next section using the results of Section 5.1, are excellent for the cases considered.

5.3. Four Approximations to the Exact Significance Level: How Do They Compare?

The decision to accept or reject a given null model for the cell probabilities depends on the calculation of a critical value c_α, for which there is only a small probability α (e.g., 0.05 or 0.01) that the test statistic will exceed this value if the null model is true. For example, consider carrying out a test of H_0: $\boldsymbol{\pi} = \boldsymbol{\pi}_0$ using the test statistic $2nI^\lambda(\mathbf{x}/n : \boldsymbol{\pi}_0)$, where $\mathbf{x}$ is the observed value of the multinomial random vector $\mathbf{X}$. Due to the discrete nature of the power-divergence statistic, we need to find a constant c_α so that, for a given λ and an $\alpha 100\%$ significance level,

$$\begin{aligned} Pr(2nI^\lambda(\mathbf{X}/n : \boldsymbol{\pi}_0) \geq c_\alpha | H_0) &\geq \alpha \\ Pr(2nI^\lambda(\mathbf{X}/n : \boldsymbol{\pi}_0) > c_\alpha | H_0) &< \alpha. \end{aligned} \tag{5.11}$$

We can write this equivalently as

$$\begin{aligned} T_E(c_\alpha) &\geq \alpha \\ T_E(c_\alpha + \varepsilon) &< \alpha; \qquad \varepsilon > 0, \end{aligned} \tag{5.12}$$

where $T_E(c) = Pr(2nI^\lambda(\mathbf{X}/n : \boldsymbol{\pi}_0) \geq c | H_0)$ for $c \geq 0$, is the exact distribution tail function of the test statistic $2nI^\lambda(\mathbf{X}/n : \boldsymbol{\pi}_0)$ assuming H_0 is true.

The literature contains many enumeration and simulation studies concerning the accuracy of using the chi-squared distribution tail function T_χ of (5.6) as an approximation to T_E for Pearson's X^2 statistic ($\lambda = 1$) and the loglikelihood ratio statistic G^2 ($\lambda = 0$) (e.g., Good et al., 1970; Roscoe and Byars, 1971; Tate and Hyer, 1973; Margolin and Light, 1974; Radlow and Alf, 1975; Chapman, 1976; Larntz, 1978; Kotze and Gokhale, 1980; Lawal, 1984; Kallenberg et al., 1985; Hosmane, 1986; Koehler, 1986; Rudas, 1986). Much of what follows in this section generalizes these studies and provides new insight into the accuracy for these two well-known statistics through the perspective of the power-divergence family. On the other hand, the adequacy of the normal approximation of Section 4.3 has been discussed very little previously (Lawal, 1980; Koehler and Larntz, 1980; Koehler, 1986). Although we have provided some initial research in this area, further studies are needed (Section 8.1).

Now let us consider the four approximations to $T_E(c)$ defined so far. For a given λ these are:

(a) $T_\chi(c)$—the chi-squared distribution tail function (5.6);
(b) $T_C(c)$—the moment-corrected chi-squared distribution tail function (5.7);
(c) $T_S(c)$—the second-order-corrected chi-squared distribution tail function (5.8); and
(d) $T_N(c)$—defined as $Pr(N(0,1) \geq (c - \mu_k^{(\lambda)})/\sigma_k^{(\lambda)}))$, the tail function of the normal distribution of Section 4.3, where $N(0,1)$ is a standard normal random variable and $\mu_k^{(\lambda)}$, $\sigma_k^{(\lambda)}$ are given by (4.16).

Accuracy When the Model Is Completely Specified

For a completely specified null model and given values of n, k, and λ, the exact distribution tail function T_E for $2nI^\lambda(\mathbf{X}/n : \boldsymbol{\pi}_0)$ can be calculated by enumerating all possible combinations $\mathbf{x} = (x_1, x_2, \ldots, x_k)$ of n observations classified into k cells. We obtain $T_E(c)$ by selecting all of those $\mathbf{x}$ for which $2nI^\lambda(\mathbf{x}/n : \boldsymbol{\pi}_0)$ is greater than or equal to the specified critical value c, and summing their respective multinomial probabilities.

Most of the small-sample studies published in the literature have concentrated on the equiprobable hypothesis (5.5). The reasons for this are: (a) equiprobable class intervals produce the most sensitive tests (Section 3 of the Historical Perspective); (b) by applying the probability integral transformation, many goodness-of-fit problems reduce to testing the fit of the uniform distribution on [0, 1] (Section 7.1); and (c) the calculations for T_E are greatly reduced (Read, 1984b). Consequently we shall concentrate on the equiprobable hypothesis in this section. Furthermore we shall choose the critical value c to be the chi-squared critical value of size α (i.e., $c = \chi^2_{k-1}(\alpha)$ and hence $T_\chi(c) = \alpha$) because this is the approximation most frequently used in practice, and it does not depend on λ. Read (1984b) compares the magnitudes of $|T_\chi(\chi^2_{k-1}(\alpha)) - T_E(\chi^2_{k-1}(\alpha))|$ for values of k from 2 to 6 and values of n from 10 to 50, and shows that for $10 \leq n \leq 20$ the chi-squared approximation $T_\chi(\chi^2_{k-1}(\alpha)) = \alpha$ is accurate for $T_E(\chi^2_{k-1}(\alpha))$ provided $\lambda \in [1/3, 3/2]$. All these cases satisfy the minimum expected cell size criterion, $\min_{1 \leq i \leq k} n\pi_i \geq 1$. These results for general λ are consistent with Larntz (1978), who considers some more general hypotheses but only for X^2 ($\lambda = 1$) and G^2 ($\lambda = 0$). He concludes that "the Pearson statistic appears to achieve the desired [chi-squared significance] level in general when all cell expected values are greater than 1.0" (Larntz, 1978, p. 256).

Figures 5.1 and 5.2 illustrate values of the four approximations together with the exact significance level for $n = 20$, $k = 5$ and $n = 20$, $k = 6$ respectively, where α is set to 0.1. The poor accuracy of the normal approximation is obvious from these figures and is noted by Read (1984b) for many values of n and k. As n becomes larger, there is a range of λ for which we can use the chi-squared critical value to approximate the exact value. However as k increases for fixed n, the error in the significance level increases for tests using λ outside the interval [1/3, 3/2] (e.g., consider Figure 5.1 versus Figure 5.2).

Assuming the equiprobable hypothesis and the combinations of n, k

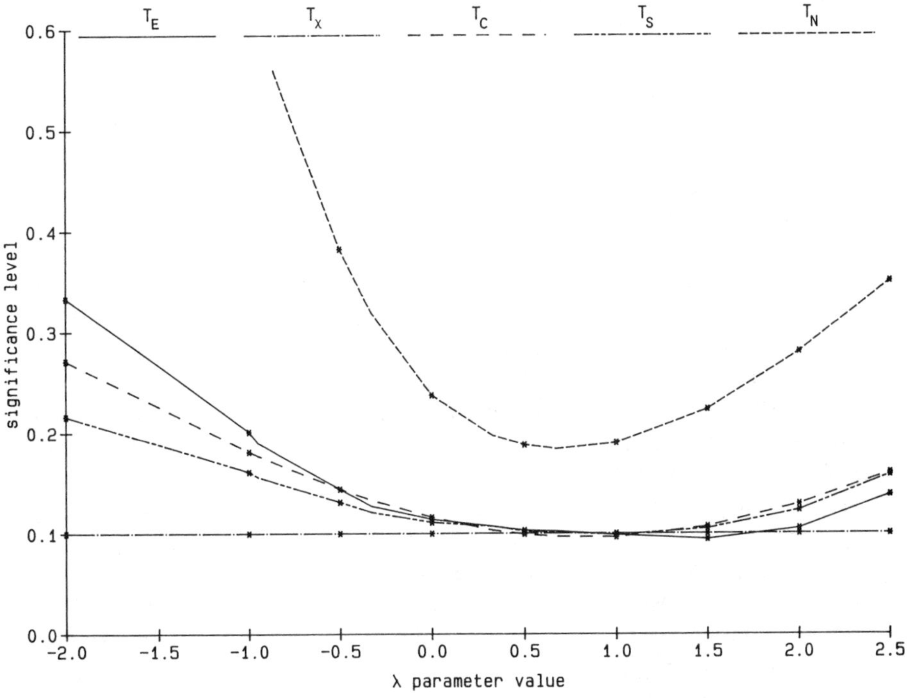

Figure 5.1. Comparison of the four approximations to $T_E(c)$ assuming the equiprobable hypothesis. $c = \chi^2_{k-1}(\alpha)$; $\alpha = 0.1$; $n = 20$; $k = 5$. [Source: Read (1984b)]

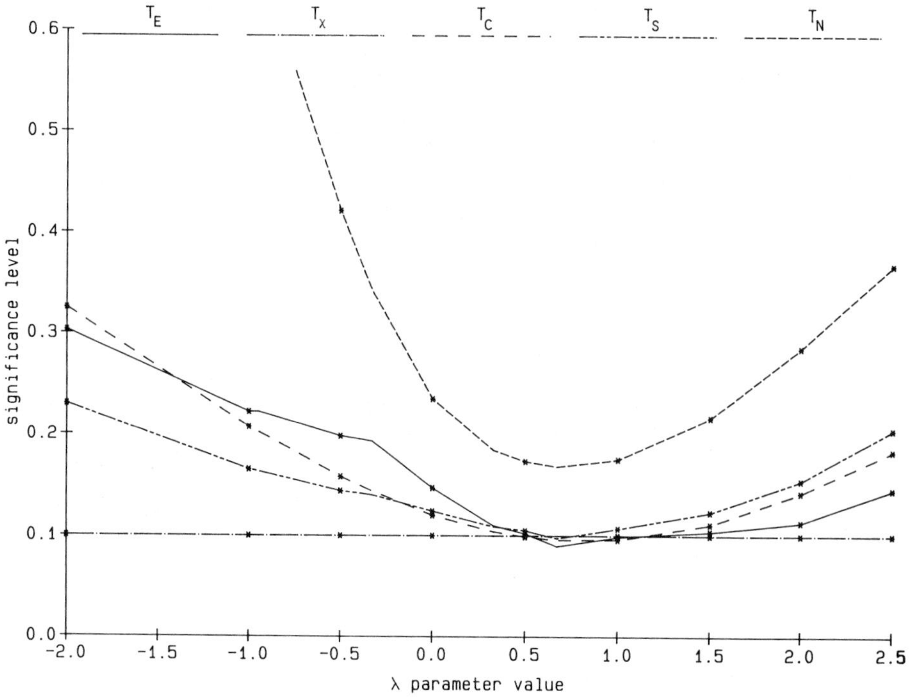

Figure 5.2. Comparison of the four approximations to $T_E(c)$ assuming the equiprobable hypothesis. $c = \chi^2_{k-1}(\alpha)$; $\alpha = 0.1$; $n = 20$; $k = 6$. [Source: Read (1982)]

and λ considered here, the moment-corrected chi-squared approximation $T_C(\chi^2_{k-1}(\alpha))$ is generally a more accurate approximation for $T_E(\chi^2_{k-1}(\alpha))$ than is the uncorrected chi-squared approximation $T_\chi(\chi^2_{k-1}(\alpha))$ (Figures 5.1 and 5.2; Read, 1984b). However, the results of Section 5.1 show that $T_C(\chi^2_{k-1}(\alpha))$ and $T_\chi(\chi^2_{k-1}(\alpha))$ will have similar values for $\lambda \in [2/3, 1]$ (Table 5.1); consequently $T_\chi(\chi^2_{k-1}(\alpha))$ will be closest to the exact distribution tail function $T_E(\chi^2_{k-1}(\alpha))$ for $\lambda \in [2/3, 1]$.

Combining the results of this section together with Section 5.1 and the recommendations on minimum expected cell size for Pearson's X^2 due to Larntz (1978) and Fienberg (1980, p. 172), we conclude that the traditional chi-squared critical value $\chi^2_{k-1}(\alpha)$ can be used with accuracy for general k (when testing the equiprobable hypothesis) provided $\min_{1 \leq i \leq k} n\pi_i \geq 1$, and $\lambda \in [2/3, 1]$.

For the simple hypothesis (i.e., no parameter estimation) with unequal cell probabilities, the results of Section 5.1 suggest choosing $\lambda \in [0.61, 0.67]$. This recommendation requires further verification based on small-sample studies. (Relevant small-sample studies when parameters are estimated are discussed later in this section and in Section 8.1.)

Clearly the power-divergence statistic with $\lambda = 2/3$ fares well throughout this discussion, X^2 ($\lambda = 1$) also does well, but the significance levels for G^2

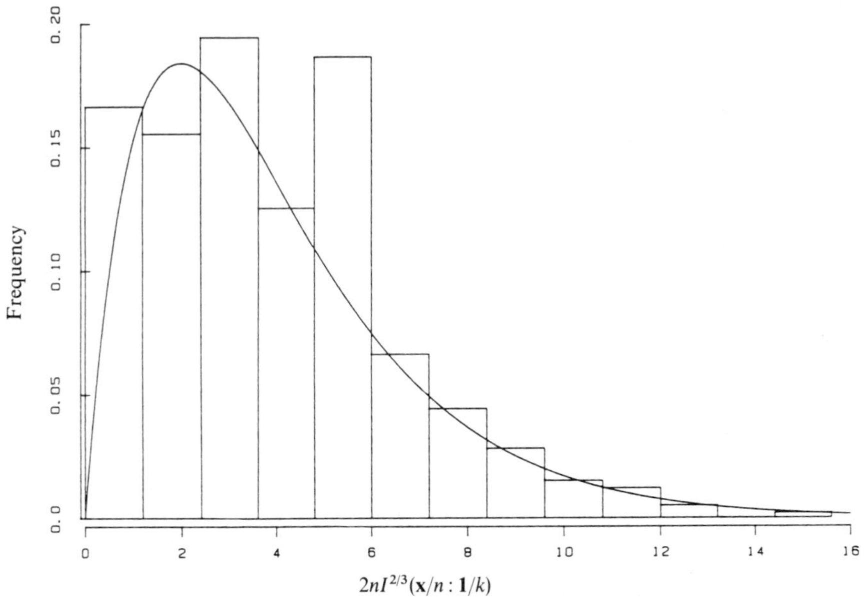

Figure 5.3. Histogram of the exact distribution function for the power-divergence statistic with $\lambda = 2/3$, assuming the equiprobable hypothesis, with $n = 10$ and $k = 5$. The chi-squared distribution function with four degrees of freedom is given by the overlaid curve. [Source: Read (1984b)]

($\lambda = 0$) and the Freeman-Tukey statistic F^2 ($\lambda = -1/2$) are not so well approximated by the chi-squared significance level. This conclusion is further verified by the studies of Margolin and Light (1974), Chapman (1976), Larntz (1978), and Kallenberg et al. (1985) and is explained in Section 6.3. To illustrate how well the statistic with $\lambda = 2/3$ approximates a chi-squared random variable for the equiprobable hypothesis, Figure 5.3 overlays a chi-squared distribution onto the exact distribution for $2nI^{2/3}(\mathbf{X}/n : \mathbf{1}/k)$ with $n = 10$ and $k = 5$. The approximation is excellent in the right-hand tail, precisely where it needs to be for calculating significance levels.

Accuracy When Parameters Are Estimated

Much less work has been done on assessing the accuracy of the chi-squared approximation when the null model requires parameters to be estimated; for example, in the case of loglinear models for contingency tables. Unfortunately, the exact distribution function will depend on the unknown parameters in the model (generally called *nuisance parameters*) and cannot be evaluated. A common solution to this problem is to condition on sufficient statistics for these nuisance parameters (e.g., Cochran, 1952; McCullagh, 1985b, 1986;

Section 8.1), so the resulting conditional distribution no longer depends on the unknown parameters. We wish to emphasize that although this is a common-sense way around a difficult problem, it is ad hoc, and does not claim to be the only possible approach (e.g., Koehler, 1986).

As an example, consider an $r \times c$ contingency table with multinomial cell frequencies X_{ij}; $i = 1, \ldots, r; j = 1, \ldots, c$, giving

$$Pr(\mathbf{X} = \mathbf{x}) = n! \prod_{i=1}^{r} \prod_{j=1}^{c} \pi_{ij}^{x_{ij}}/x_{ij}! \tag{5.13}$$

where $n = \sum_{i=1}^{r} \sum_{j=1}^{c} x_{ij}$. From Section 3.2 recall the null model for independence between rows and columns is

$$H_0\colon \pi_{ij} = \pi_{i+}\pi_{+j}, \tag{5.14}$$

and (5.13) becomes

$$Pr(\mathbf{X} = \mathbf{x}) = \left[n! \prod_{i=1}^{r} \frac{\pi_{i+}^{x_{i+}}}{x_{i+}!}\right]\left[n! \prod_{j=1}^{c} \frac{\pi_{+j}^{x_{+j}}}{x_{+j}!}\right]\left[\frac{\prod_{i=1}^{r} x_{i+}! \prod_{j=1}^{c} x_{+j}!}{n! \prod_{i=1}^{r} \prod_{j=1}^{c} x_{ij}!}\right],$$

which depends on the unknown nuisance parameters $\pi_{i+}; i = 1, \ldots, r$ and π_{+j}; $j = 1, \ldots, c$. The first (second) bracketed term gives the probabilities of the row (column) totals, and the third term gives the probabilities of the cell frequencies for fixed values of the row and column totals. Consequently, conditioning on the row and column totals, we obtain

$$Pr(\mathbf{X} = \mathbf{x} | x_{1+}, \ldots, x_{r+}, x_{+1}, \ldots, x_{+c}) = \left[\frac{\prod_{i=1}^{r} x_{i+}! \prod_{j=1}^{c} x_{+j}!}{n! \prod_{i=1}^{r} \prod_{j=1}^{c} x_{ij}!}\right], \tag{5.15}$$

which is independent of $\pi_{i+}; i = 1, \ldots, r$ and $\pi_{+j}; j = 1, \ldots, c$, provided the null model (5.14) is correct. The case $r = c = 2$ is the well-known Fisher's exact test for the 2×2 contingency table, and (5.15) simplifies to the hypergeometric distribution.

There are various algorithms available for computing all $r \times c$ tables with their associated probabilities conditional on row and column totals (e.g., Hancock, 1975; Baker, 1977). Mehta and Patel (1986) present an algorithm that calculates only those tables whose probabilities are less than the probability of the observed table; hence it is very efficient for large tables. On the other hand, the algorithm of Agresti et al. (1979) takes a sample from the set of all possible tables and is useful for approximating the attained significance level for very large tables. Verbeek and Kroonenberg (1985) provide a useful survey of these algorithms and many others.

The published small-sample studies concentrate on testing independence in two-dimensional tables and use (5.15) to evaluate the exact (conditional) distribution of X^2 ($\lambda = 1$) and/or G^2 ($\lambda = 0$) (e.g., Margolin and Light, 1974; Larntz, 1978; Lawal and Upton, 1984; Hosmane, 1986). Haldane (1939) and Lewis et al. (1984) derive correction factors for the first three conditional moments of X^2 for testing independence in an $r \times c$ contingency table. The current comparative studies of X^2 and G^2 tend to support our conclusions that $\lambda = 1$ is preferable to $\lambda = 0$ in terms of the accuracy of the chi-squared significance level (Section 1 of the Historical Perspective).

Recently, Rudas (1986) and Hosmane (1987) have shown that the statistic based on $\lambda = 2/3$ has a significance level closely approximated by the chi-squared distribution tail function for testing independence in two- and some three-dimensional tables, and for testing no three-factor interaction in three-dimensional tables. Their simulations indicate that using $\lambda = 2/3$ produces a test statistic that is similar to X^2 in terms of the small-sample accuracy of the chi-squared significance levels; however G^2 does not do so well and tends to reject the null hypothesis too frequently in the cases considered. These results support our findings for cases where no parameters are estimated and the minimum expected cell frequency is no less than 1. We conjecture that further studies with parameter estimation will confirm that the power-divergence statistic with $\lambda = 2/3$ is an excellent compromise between X^2 ($\lambda = 1$) and G^2 ($\lambda = 0$). Further discussion on the effect of parameter estimation is contained in Section 8.1.

Summary

In summary, the critical value for the power-divergence statistic with $\lambda = 2/3$ is well approximated by the chi-squared critical value when testing the equiprobable hypothesis (5.5), provided $n/k \geq 1$ and $n \geq 10$. Pearson's X^2 ($\lambda = 1$) also does well in this situation. For the other well-known statistics (e.g., G^2, loglikelihood ratio ($\lambda = 0$); F^2, Freeman-Tukey ($\lambda = -1/2$); NM^2, Neyman-modified X^2 ($\lambda = -2$)) a correction factor is needed when $n \leq 20$; we recommend using a correction based on T_C that is relatively easy to calculate. The studies of Koehler and Larntz (1980), Rudas (1986), and Hosmane (1987) offer preliminary evidence that the chi-squared approximation can be accurate for $\lambda = 2/3$ and $\lambda = 1$ when the expected values are as low as 1/4 (provided $n^2/k \geq 10$, $n \geq 10$, and $k \geq 3$).

The accuracy of the chi-squared critical value for the power-divergence statistic based on $\lambda = 2/3$ and $\lambda = 1$ appears to carry over to hypotheses with unequal cell probabilities and estimated parameters, provided $\min_{1 \leq i \leq k} n\pi_i \geq 1$ (however if some of the expectations are very small while others are greater than 1 there are no firm recommendations regarding the best approximation to use; see, e.g., Koehler, 1986). Preliminary results involving parameter estimation lead us to conjecture similar accuracy for these two statistics under loglinear models (Section 8.1). Section 6.3 offers further insight into the reasons behind these results.

5.4. Exact Power Comparisons

In Sections 4.2 and 4.3, we considered a variety of asymptotic-efficiency criteria depending on the type of asymptotics assumed (classical (fixed-cells) assumptions or sparseness assumptions) and on the type of alternative models assumed (nonlocal (fixed) alternatives or local alternatives converging to the null). Our conclusions were very sensitive to both the asymptotics and the alternatives.

Now we are interested in investigating the question of the small-sample efficiency of the power-divergence statistic. One approach to this problem is to find conditions under which the asymptotic results of Sections 4.2 and 4.3 are approximately correct in small samples. However, this approach is complicated, and Frosini (1976) notes that for Pearson's X^2 statistic the conditions under which the noncentral chi-squared results of Section 4.2 are appropriate are very restrictive in small samples.

A more direct approach, which we pursue here, is to calculate the exact power of $2nI^{\lambda}(\mathbf{X}/n : \boldsymbol{\pi}_0)$ for each λ, and to compare the values at specific alternatives without any reference to asymptotics. This approach is similar to one used by West and Kempthorne (1971) and Haber (1984) to compare the power of X^2 and G^2. For the same reasons as outlined in Section 5.3, we restrict our attention to the equiprobable null model (5.5) and consider alternative models where one of the k probabilities is perturbed, and the rest are adjusted so that they still sum to 1:

$$H_1: \pi_i = \begin{cases} (1 - \delta/(k-1))/k; & i = 1, \ldots, k-1 \\ (1+\delta)/k; & i = k, \end{cases} \tag{5.16}$$

where $-1 \leq \delta \leq k-1$.

The Randomized Size-α Test

To produce power calculations for each test statistic $2nI^{\lambda}(\mathbf{X}/n : \boldsymbol{\pi}_0)$, it is necessary to choose a significance level $\alpha 100\%$ and calculate the associated critical value c_α. This produces a *size-α test.* The naive approach is to rely on one of the approximations defined in Section 5.3; however we observed that the magnitude of the approximation error depends on λ. Such over- or underestimation of the exact size of the tests will confound any power comparisons with λ-dependent approximation errors. A better approach is to calculate the critical value of the exact size-α test using the results of Section 5.3. However due to the discrete nature of the critical values, the attainable significance levels for each test statistic also will be discrete. Therefore it is unlikely, for any given α, that an exact size-α test exists for every λ. To overcome this problem we use the appropriate randomized size-α test, which can always be obtained regardless of λ, thus making the power functions of each test directly comparable across λ values.

The *randomized size-α test* is defined as follows: Let $c_\lambda(\alpha)$ be an achievable value of $2nI^\lambda(\mathbf{X}/n:\boldsymbol{\pi}_0)$ so that

$$\begin{aligned} Pr(2nI^\lambda(\mathbf{X}/n:\boldsymbol{\pi}_0) > c_\lambda(\alpha)|H_0) &= \alpha_{1,\lambda} \\ Pr(2nI^\lambda(\mathbf{X}/n:\boldsymbol{\pi}_0) \geq c_\lambda(\alpha)|H_0) &= \alpha_{2,\lambda}, \end{aligned} \tag{5.17}$$

with $\alpha_{1,\lambda} < \alpha \leq \alpha_{2,\lambda}$. Given an observed outcome $\mathbf{x}$, the randomized size-α test rejects H_0 with probability:

$$\begin{cases} 1, & \text{if } 2nI^\lambda(\mathbf{x}/n:\boldsymbol{\pi}_0) > c_\lambda(\alpha) \\ \dfrac{\alpha - \alpha_{1,\lambda}}{\alpha_{2,\lambda} - \alpha_{1,\lambda}}, & \text{if } 2nI^\lambda(\mathbf{x}/n:\boldsymbol{\pi}_0) = c_\lambda(\alpha) \\ 0, & \text{if } 2nI^\lambda(\mathbf{x}/n:\boldsymbol{\pi}_0) < c_\lambda(\alpha). \end{cases}$$

It is easily verified from (5.17) that the exact size of this test is

$$1 \cdot \alpha_{1,\lambda} + \frac{\alpha - \alpha_{1,\lambda}}{\alpha_{2,\lambda} - \alpha_{1,\lambda}} \cdot (\alpha_{2,\lambda} - \alpha_{1,\lambda}) = \alpha,$$

as required. Defining $\beta_{1,\lambda}$ and $\beta_{2,\lambda}$ as

$$\begin{aligned} Pr(2nI^\lambda(\mathbf{X}/n:\boldsymbol{\pi}_0) > c_\lambda(\alpha)|H_1) &= \beta_{1,\lambda} \\ Pr(2nI^\lambda(\mathbf{X}/n:\boldsymbol{\pi}_0) \geq c_\lambda(\alpha)|H_1) &= \beta_{2,\lambda}, \end{aligned}$$

for H_1 defined in (5.16), we obtain the power of the randomized size-α test to be

$$\beta_\lambda = 1 \cdot \beta_{1,\lambda} + \frac{\alpha - \alpha_{1,\lambda}}{\alpha_{2,\lambda} - \alpha_{1,\lambda}} \cdot (\beta_{2,\lambda} - \beta_{1,\lambda}).$$

Exact Results for the Power-Divergence Statistic

To represent the two extremes of the alternative model (5.16) together with a midpoint, we consider three values of the perturbation factor δ: $\delta = 1.5$, $\delta = 0.5$, and $\delta = -0.9$. The exact power of the randomized size-α test with $\alpha = 0.05$ is summarized by Read (1984b) for values of k from 2 to 6 and n from 10 to 50, and we paraphrase these results here.

For the case $k = 2$, the power-divergence statistic has identical critical values and power functions for every value of $\lambda \in [-5, 5]$. For $k > 2$, the power function increases with λ for $\delta > 0$ (which represents a *bump* alternative) but decreases with λ for $\delta < 0$ (which represents a *dip* alternative). However for $|\lambda|$ that is large, the rate of change in the power function is very small, indicating a *plateau* effect of the power with respect to λ. Table 5.2 illustrates these observations when $n = 20$ and $k = 5$. Wakimoto et al. (1987) use the same alternatives (5.16), and reach the same conclusions in the special cases X^2 ($\lambda = 1$), G^2 ($\lambda = 0$), F^2 ($\lambda = -1/2$) by plotting the power against δ for $k = 5$, $n = 30$ and $k = 6$, $n = 20$ (both with $\alpha = 0.05$).

Table 5.2. Exact Power Functions of the Randomized Test for the Equiprobable Null Model (5.5) Against the Alternative (5.16)

$\alpha = 0.05, n = 20, k = 5$			
	δ		
λ	1.5	0.5	−0.9
−5.0	0.2253	0.0742	0.5880
−2.0	0.2253	0.0742	0.5880
−1.0	0.2253	0.0742	0.5880
−0.5	0.3361	0.0800	0.5875
−0.3	0.4468	0.0864	0.5693
0.0	0.6100	0.1073	0.4466
0.3	0.6605	0.1146	0.3760
0.5	0.6806	0.1188	0.3216
0.7	0.6907	0.1211	0.2851
1.0	0.6997	0.1228	0.2720
1.5	0.7168	0.1257	0.2297
2.0	0.7306	0.1278	0.1896
2.5	0.7392	0.1288	0.1760
5.0	0.7498	0.1296	0.1464

Source: Read (1984b).

In summary, we see that for detecting bump type alternatives, we should choose λ as large as other restrictions will allow, but for dip type alternatives, choosing λ large and negative will improve the power. However, we note that for small sample size n relative to the number of cells k, there is a high chance of an individual cell having a zero observed frequency. In this situation the power-divergence statistic with $\lambda \le -1$ will be undefined, therefore we recommend choosing $\lambda > -1$ (as in Section 4.3).

Other Studies

These observations coincide with the power comparisons of West and Kempthorne (1971) and Koehler and Larntz (1980), who consider X^2 ($\lambda = 1$) and G^2 ($\lambda = 0$). For the equiprobable null hypothesis, Koehler and Larntz (1980, p. 340) state "The Pearson test is decidedly dominant as the alternative moves toward a boundary of [the probability simplex they define by] T_k that contains a high proportion of zeros and a few relatively large probabilities. The likelihood ratio test is dominant at alternatives that lie near boundaries of T_k that contain a small proportion of near-zero probabilities and have nearly equal probabilities in the remaining cells." The bump alternatives $\delta = 1.5$ and $\delta = 0.5$ lie near the first boundary mentioned by Koehler and Larntz, whereas the dip alternative $\delta = -0.9$ lies near their second boundary. Therefore if we constrain λ to the two-point set $\{0, 1\}$, as Koehler and Larntz do, then our recommendations are equivalent to theirs. However, our study looks at the traditional statistics as part of a continuum, so we can improve on Koehler and Larntz's results by allowing λ to vary further as discussed earlier. For example, if we wish to choose a conservative test that has reasonable power against both bump and dip alternatives, Read (1984b) concludes that we should choose $\lambda \in [1/3, 2/3]$.

For the hypothesis of no three-factor interaction in three-dimensional contingency tables, Hosmane (1987) compares by simulation, the members $\lambda = -1/2, 0, 2/3$, and 1 of the power-divergence family. He finds the attained significance levels are quite close to the chi-squared nominal values for $\lambda = 2/3$ and $\lambda = 1$, and proceeds to simulate the power of these two statistics. Hosmane's table VI indicates that the power-divergence statistic with $\lambda = 2/3$ is as efficient or more efficient than Pearson's X^2 ($\lambda = 1$) 40 out of 48 times when $\alpha = 0.05$, and 34 out of 48 times when $\alpha = 0.01$.

In the case of null models with unequal cell probabilities, the results of Chapter 6 indicate that the power of the test statistics will increase with λ for alternatives that have a large ratio of the alternative/null expected frequency in one cell (this is a generalization of the bump alternatives to the null model with unequal probabilities). Conversely, the power will decrease with λ for alternatives that have a near-zero ratio of the alternative/null expected frequency in one cell (generalizing the dip alternatives). These conclusions are confirmed by the results of Kallenberg et al. (1985) using heavy-tailed and light-tailed alternatives. Drost et al. (1987) make the same recommendations from a study of two new approximations to the power of the test based on the power-divergence statistic (Section 4.2).

An interesting discussion of the general spike (i.e., bump) detection problem for multinomials is provided by Levin (1983). Among other things, he derives the distribution of the maximum cell frequency (i.e., the power-divergence statistic with $\lambda \to \infty$) and compares this statistic graphically with Pearson's X^2 for detecting the bump type alternatives from (5.16).

5.5. Which Test Statistic?

With so many recommendations available, how can we choose a test statistic (i.e., a value of λ) that will best suit our requirements? To answer this question we need to consider the objectives for testing the null model.

If we are looking for a test statistic with an easily calculable critical value and that provides reasonable protection against the extreme bump and dip alternative models, then we recommend using $\lambda = 2/3$. For small n, we believe this to be the best choice of λ. For large n, it is an excellent compromise between the loglikelihood ratio statistic G^2 ($\lambda = 0$), which is optimal for nonlocal (fixed) alternatives (using Bahadur efficiency), and the Pearson X^2

statistic ($\lambda = 1$), which is optimal under the sparseness assumptions of Section 4.3 (using Pitman efficiency and assuming the equiprobable null model). We recommend that the critical value be calculated from the chi-squared approximation provided $n \geq 10$ and the minimum expected cell frequency is no less than 1 (when the expected frequencies are all nearly equal, the chi-squared approximation may be valid for expectations as low as 1/4; see Section 5.3). If some of the expected cell frequencies are very small while others are greater than 1, then there are no firm recommendations. The adequacy of the normal approximation for finite values of n needs further investigation (Section 8.1).

If we are looking for a statistic to guard against a specific type of extreme alternative, and we are willing to use one of the more complicated approximations (described in this chapter) to find the critical value, then the results of Sections 4.2, 4.3, 5.4, and 6.4 give some insight into choosing λ. That is, use larger values of λ to detect a departure from the null model that involves one cell obtaining a large ratio of the alternative/null expected frequency. Otherwise, use larger negative values of λ to detect a departure from the null model that involves one cell obtaining a near-zero ratio of the alternative/null expected frequency.

There appears to be little point in using $|\lambda|$ much larger than five, because (a) in the cases we have examined, the gain in power becomes very small; and (b) the use of approximations to obtain the critical value for rejecting the null model becomes more suspect. Even for sample sizes as large as 30, we observed that for $k = 3$ and a 1% nominal significance level (i.e., a $\chi_2^2(0.01)$ critical value), the exact significance level is 6% for $\lambda = 5$ and 11% for $\lambda = -5$. For larger k the approximation error is worse as noted in Section 5.3. Consequently for λ-values outside the interval $[1/3, 3/2]$ we recommend that a moment-corrected approximation based on (5.7) should be used whenever $n \geq 10$ and the minimum expected cell frequency is no less than 1. (Kallenberg et al., 1985, provide further discussion for the case $\lambda = 0$.)

Finally, because the existence of zero counts causes the power-divergence statistic to be undefined for $\lambda \leq -1$, we recommend using $\lambda > -1$ for large sparse arrays.

The enumerative results of this chapter are for the equiprobable null model with completely specified alternatives. We have indicated where we believe similar results will carry through for more complex models that may involve estimated parameters. However, more enumeration and simulation studies are needed to fully substantiate these generalizations. Such studies are sparse in the literature and tend to be very specific due to the number of variables and assumptions that need to be taken into account (Section 1 of the Historical Perspective). These include the sample size n, the number of cells k, the form of null model (loglinear, etc.), and the "direction of departure" of the alternative from the null.

CHAPTER 6

Comparing the Sensitivity of the Test Statistics

Chapters 4 and 5 presented some theoretical and computational comparisons of the power-divergence family members. This chapter provides some further practical understanding of these results, by analyzing the effects of individual cell frequencies on the power-divergence statistic.

Sections 6.1 and 6.2 provide the background and results on individual cell contributions necessary for making comparisons between family members. In Section 6.3 we use these results to explain the conclusions about exact-test significance levels and power obtained in Chapter 5. Three illustrations of the effects of individual cell contributions on the power-divergence statistic are provided in Section 6.4. Section 6.5 presents a different perspective of the power transformation parameter λ by considering the distribution of an individual transformed cell frequency when the expected frequency is small. In Section 6.6 a geometric interpretation of this power transformation provides a partial explanation for the good performance of the power-divergence statistic with $\lambda = 2/3$. Finally Section 6.7 gives recommendations on which values of λ to use.

6.1. Relative Deviations between Observed and Expected Cell Frequencies

Chapters 2 through 5 illustrated three important characteristics of the power-divergence statistic as a function of varying λ. These are:

(a) The power-divergence statistic, defined from (2.17) as

$$2I^{\lambda}(\mathbf{x}:\mathbf{m}) = \frac{2}{\lambda(\lambda+1)} \sum_{i=1}^{k} x_i \left[\left(\frac{x_i}{m_i} \right)^{\lambda} - 1 \right],$$

decreases and then starts to increase with increasing values of λ (Tables 2.3, 3.3, and 3.5). The value of λ at which the minimum occurs depends on x_i and m_i, where x_i is the observed frequency and m_i is the expected frequency for cell i under a hypothesized model and $i = 1, \ldots, k$.

(b) For λ outside the interval $[1/3, 3/2]$, the traditional chi-squared significance level tends to underestimate the exact significance level of the test based on the power-divergence statistic when $\min_{1 \le i \le k} m_i \ge 1$. Furthermore, as $|\lambda|$ increases in magnitude, so does the discrepancy between these two levels (Section 5.3).

(c) For bump type alternative hypotheses, the power of the test increases with λ; whereas for dip type alternatives, the power decreases with increasing λ (Section 5.4).

The objective of this chapter is to explain these three different effects in terms of the functional sensitivity of the power-divergence statistic to the relative size of the observed and expected cell frequencies. The motivation behind looking for explanations in this direction comes from the work of Larntz (1978), who compares the exact versus chi-squared significance levels for three members of the power-divergence family (see also Sections 1 and 2 of the Historical Perspective). Specifically, he compares:

Pearson's X^2 statistic ($\lambda = 1$)

$$X^2 = 2I^1(\mathbf{x} : \mathbf{m}) = \sum_{i=1}^{k} (x_i - m_i)^2/m_i, \tag{6.1}$$

the loglikelihood ratio statistic ($\lambda = 0$)

$$G^2 = 2I^0(\mathbf{x} : \mathbf{m}) = 2 \sum_{i=1}^{k} x_i \log(x_i/m_i), \tag{6.2}$$

and a variant of the Freeman-Tukey statistic F^2 ($\lambda = -1/2$) from (2.11), defined as

$$T^2 = \sum_{i=1}^{k} (\sqrt{x_i} + \sqrt{x_i + 1} - \sqrt{4(m_i + 1)})^2. \tag{6.3}$$

In fact T^2 is asymptotically equivalent to F^2 (Bishop et al., 1975) and is sometimes referred to as the modified Freeman-Tukey statistic (Fienberg, 1979; Lawal and Upton, 1980; Lawal, 1984).

The results for the equiprobable null hypothesis in Section 5.3 (where the expected cell frequencies are greater than 1) indicate that tests based on the power-divergence statistic with $\lambda \in [1/3, 3/2]$ achieve levels close to the chi-squared significance level. However for λ outside this range, the exact test level is much higher than the chi-squared level. Margolin and Light (1974) and Larntz (1978) indicate that both G^2 ($\lambda = 0 \notin [1/3, 3/2]$) and T^2 (asymptotically, $\lambda = -1/2 \notin [1/3, 3/2]$) tend to reject too frequently for expected cell frequencies in the range 1.5 to 4; on the other hand X^2 ($\lambda = 1 \in [1/3, 3/2]$) tends to achieve the desired level, provided all expected frequencies are greater than 1. Therefore these earlier results support our conclusion.

To explain these discrepancies in behavior for varying λ, we consider the sensitivity of the power-divergence statistic to a single observed frequency, as a function of λ. Section 6.2 provides a general formula for the minimum asymptotic contribution from an individual cell of the multinomial distribution. This formula generalizes the results of Larntz (1978) for X^2, G^2, and T^2 to the power-divergence family, and is used in Section 6.3 to explain effects (b) and (c) described earlier. Effect (a) is described further in Section 6.4, where we illustrate the sensitivity of $2I^\lambda(\mathbf{x}:\mathbf{m})$ to the ratios x_i/m_i for $i = 1, \ldots, k$ as λ changes.

6.2. Minimum Magnitude of the Power-Divergence Test Statistic

Throughout this section, we assume that the observed frequency in cell j (j fixed) is $x_j = \delta$, where $0 \le \delta < n$ is fixed. We now calculate the minimum value possible for $2I^\lambda(\mathbf{x}:\mathbf{m})$ where $\mathbf{x}$ and $\mathbf{m}$ are the observed and expected frequency vectors, respectively. Assume that the other x_is are continuous variables on $[0, n-\delta]$ with $\sum_{i=1}^k x_i = n$, and that $m_i > 0$ for all $i = 1, \ldots, k$. Then it is straightforward to show that $2I^\lambda(\mathbf{x}:\mathbf{m})$ will be minimized when the x_is ($i \ne j$) are spread evenly over the cells in proportion to the expected frequencies; that is, x_i/m_i is constant for all i ($i \ne j$). Consequently the following result is proved in Section A13 of the Appendix. When $x_j = \delta$ is fixed, the minimum asymptotic value of $2I^\lambda(\mathbf{x}:\mathbf{m})$ is

$$\frac{2}{\lambda(\lambda+1)}\left\{\delta\left[\left(\frac{\delta}{m_j}\right)^\lambda - 1\right] + \lambda(m_j - \delta)\right\}; \qquad \lambda \ne 0, \lambda \ne -1, \tag{6.4}$$

where m_j is the expected cell frequency associated with the cell having observed frequency $x_j = \delta$, $0 \le \delta < n$. The cases $\lambda = 0$ and $\lambda = -1$ are obtained by taking limits in (6.4) using the same argument as in Section 2.4; i.e.,

$$\begin{aligned} &2\delta\log(\delta/m_j) + (m_j - \delta), && \text{for } \lambda = 0,\\ &2m_j\log(m_j/\delta) + (\delta - m_j), && \text{for } \lambda = -1. \end{aligned} \tag{6.5}$$

To illustrate how the minimum asymptotic values (6.4) and (6.5) vary with changes in δ, m_j, and λ, we consider the two cases $\delta = 0$ and $\delta > 0$ separately.

Case 1: $\delta = 0$

First consider the effect of a single zero observed cell frequency on the minimum asymptotic value of the statistic $2I^\lambda(\mathbf{x}:\mathbf{m})$ as m_j and λ change. In this case (6.4) and (6.5) reduce to

$$\begin{aligned} &2m_j/(\lambda+1); && \lambda > -1\\ &\infty; && \lambda \le -1. \end{aligned} \tag{6.6}$$

Recall that for $\lambda \le -1$, the statistic $2I^\lambda(\mathbf{x}:\mathbf{m})$ is infinite whenever one observed cell frequency is 0 (Section 4.3). As λ increases from -1, the minimum asymptotic value decreases from ∞ to 0. In other words, smaller λ-values give more emphasis to an observed vector $\mathbf{x}$ that contains a zero frequency, $x_j = 0$. For fixed $\lambda > -1$, (6.6) is directly proportional to the expected frequency m_j in cell j. This is intuitively what we would expect, since larger values of m_j suggest a greater discrepancy between the model and the observed vector $\mathbf{x}$ with $x_j = \delta = 0$.

Case 2: $\delta > 0$

When δ is nonzero, we can divide (6.4) and (6.5) by 2δ; then writing $\xi = \delta/m_j$, (6.4) and (6.5) become

$$h^\lambda(\xi) = [\xi^\lambda - 1 + \lambda(1/\xi - 1)]/[\lambda(\lambda + 1)]; \qquad (6.7)$$

(provided the limits are taken as $\lambda \to -1$ and $\lambda \to 0$). When $\xi = 1$ (i.e., $\delta = m_j$), the function $h^\lambda(\xi)$ is identically 0. This is to be expected, since we can obtain $2I^\lambda(\mathbf{x} : \mathbf{m}) = 0$, by setting $x_i = m_i$, for all $i = 1, \ldots, k$. The exponential term in (6.7) indicates that $\xi = 1$ is a pivotal value for the behavior of $h^\lambda(\xi)$. For fixed $\xi < 1$ (i.e., $\delta < m_j$), $h^\lambda(\xi)$ is a decreasing function of λ; but for fixed $\xi > 1$ (i.e., $\delta > m_j$), $h^\lambda(\xi)$ is an increasing function of λ. These results are illustrated in Table 6.1, and are consistent with the extreme case $\delta = 0$ discussed previously.

Table 6.1. Values of $h^\lambda(\xi)$ Defined by (6.7) for Various ξ and λ

	ξ							
λ	0.1	0.5	0.9	1.0	1.1	2.0	5.0	10.0
−10.0	*	11.256	0.008	0.0	0.003	0.044	0.078	0.089
−5.0	*	1.300	0.007	0.0	0.004	0.077	0.150	0.175
−2.0	40.500	0.500	0.006	0.0	0.004	0.125	0.320	0.405
−1.0	14.026	0.386	0.006	0.0	0.004	0.153	0.478	0.670
−0.5	9.351	0.343	0.006	0.0	0.004	0.172	0.611	0.935
0.0	6.697	0.307	0.006	0.0	0.004	0.193	0.809	1.403
0.5	5.088	0.276	0.006	0.0	0.004	0.219	1.115	2.283
1.0	4.050	0.250	0.006	0.0	0.005	0.250	1.600	4.050
2.0	2.835	0.208	0.005	0.0	0.005	0.333	3.733	16.200
5.0	1.467	0.134	0.005	0.0	0.005	0.950	*	*
10.0	0.809	0.082	0.004	0.0	0.006	9.255	*	*

* Values greater than 10^3.
Source: Read (1984b).

What happens to $I^\lambda(\mathbf{x} : \mathbf{m})$ as $\lambda \to \pm\infty$? From the definition (2.17) we see that when λ is large and positive, then

$$I^\lambda(\mathbf{x} : \mathbf{m})/n \sim \frac{\left[\max\limits_{1 \le i \le k} \dfrac{x_i}{m_i}\right]^\lambda - 1}{\lambda(\lambda + 1)}. \qquad (6.8)$$

When λ is large and negative, then

$$I^{\lambda}(\mathbf{x}:\mathbf{m})/n \sim \frac{\left[\min\limits_{1\leq i\leq k} \frac{x_i}{m_i}\right]^{\lambda} - 1}{\lambda(\lambda+1)} \tag{6.9}$$

(Section A14 of the Appendix). Consequently, as $\lambda \to \infty$ (or $\lambda \to -\infty$), $2I^{\lambda}(\mathbf{x}:\mathbf{m})$ becomes increasingly dominated by the largest (or smallest) ratio of x_i/m_i.

In summary, equations (6.4) and (6.5) indicate that the departure of a cell's observed frequency from its expected frequency results in a minimum asymptotic value for the statistic $2I^{\lambda}(\mathbf{x}:\mathbf{m})$ that depends on λ. If the observed cell frequency is less than expected (i.e., $\delta = 0$; or $\delta > 0$ and $\xi < 1$), then the minimum asymptotic value of $2I^{\lambda}(\mathbf{x}:\mathbf{m})$ decreases as λ increases. However, if the observed frequency is more than expected (i.e., $\delta > 0$ and $\xi > 1$), then the minimum asymptotic value of $2I^{\lambda}(\mathbf{x}:\mathbf{m})$ increases as λ increases. For $|\lambda|$ small, the statistic is more stable against departure of the ratio $\xi = \delta/m_j$ from 1, in either direction.

Equations (6.8) and (6.9) confirm the point made in Section 5.4: Extreme values of λ are most useful for detecting deviations from an expected cell frequency in a single cell (i.e., bump or dip alternatives). Use of the maximum cell frequency to test for "spikes" is discussed by Levin (1983); he also derives the appropriate distributional results for hypothesis testing. In Section 5.3 we saw (under the equiprobable hypothesis) that the exact significance level of the test based on $2I^{\lambda}(\mathbf{x}:\mathbf{m})$ is severely underestimated by the chi-squared distribution for extreme λ-values. This is explained further in the next section. The practical implication is that we need to calculate a more accurate critical value before any tests of significance can be provided in these cases (Sections 5.1 and 5.2).

A Parallel with Diversity Indices

A parallel to the preceding discussion can be drawn from the work of Kempton (1979) on diversity measures for species abundance. Consider a vector of ordered relative abundances $\boldsymbol{\pi} = (\pi_1, \pi_2, \ldots, \pi_k)$ where $\pi_1 \geq \pi_2 \geq \cdots \geq \pi_k$, $\sum_{i=1}^{k} \pi_i = 1$, and k = total number of species in the community; that is, π_1 represents the most prevalent species and π_k the rarest species. One proposed measure of diversity (Hill, 1973) is

$$D^{\alpha}(\boldsymbol{\pi}) = \left[\sum_{i=1}^{k} \pi_i^{\alpha}\right]^{1/(1-\alpha)}; \qquad \alpha \in (-\infty, \infty).$$

Kempton notes that this measure differs in its sensitivity to rare and abundant species in the community, according to the value of the parameter α. As α increases, D^{α} becomes increasingly dependent on the most prevalent species, with $D^{\infty} = 1/\pi_1$. Conversely, as α becomes large negative, D^{α} becomes increasingly dependent on the rarest species, with $D^{-\infty} = 1/\pi_k$. Kempton suggests that α should be in the range (0, 1) so that D^{α} will be dominated by the more prevalent species. This would make the measure more stable against sampling fluctuations.

Furthermore, by formally setting $n = k$, $\mathbf{x} = \mathbf{1}$, and $\mathbf{m} = k\boldsymbol{\pi}$, the power-divergence family defines a measure of species abundance identical to one described by Patil and Taillie (1982), which Kempton (1979) discusses in some detail:

$$\frac{1}{\lambda - 1}\left[1 - \sum_{i=1}^{k} \pi_i^\lambda\right] = (\lambda - 1)^{-1}(1 - k^{1-\lambda}) - \lambda(k^{-\lambda})I^{-\lambda}(\mathbf{1} : k\boldsymbol{\pi}).$$

In Section 7.4 we discuss diversity indices in more detail, together with the related measures of directed divergence.

6.3. Further Insights into the Accuracy of Large-Sample Approximations

Based on his study of minimum contributions to goodness-of-fit statistics, Larntz (1978) concludes that for moderate expected values (i.e., in the range 1.5 to 4) the loglikelihood ratio statistic G^2 ($\lambda = 0$) and T^2 from (6.3) (asymptotically the Freeman-Tukey statistic F^2; $\lambda = -1/2$) tend to have critical values larger than those from a chi-squared distribution. He argues that this is due to the large contributions to the statistics made by cells with very small observed frequencies (i.e., 0 or 1). From (6.6) and (6.7), the result is more general. By embedding these statistics in a continuum of λ-values, we have seen that in the case $\delta < m_j$ (i.e., small ratios of observed to expected cell frequencies) the contribution of cell j to $2I^\lambda(\mathbf{x} : \mathbf{m})$ decreases as λ increases from $-1/2$ to 1. Conversely, we can add that in the case $\delta > m_j$ (i.e., large ratios of observed to expected cell frequencies) the contribution made to X^2 ($\lambda = 1$) will be larger than that made to G^2 ($\lambda = 0$) or F^2 ($\lambda = -1/2$), hence enlarging the respective exact critical value for X^2. By looking at very small expected cell frequencies (where δ/m_j will be large), the earlier work of Horn (1977) and Koehler and Larntz (1980) further supports our conclusions. For example when there are many expected frequencies less than 1, Koehler and Larntz (1980) note that for large sparse multinomials the first two moments of X^2 will be larger than those of G^2 (Sections 1 and 2 of the Historical Perspective).

Using a similar argument for the individual cell contributions, we can further interpret the enumerative results of Section 5.3 (where the expected frequencies are greater than 1) regarding approximate versus exact critical values. Recall that for the equiprobable hypothesis $m_i = n/k$; $i = 1, \ldots, k$, the error in using the traditional chi-squared critical value to approximate the exact $\alpha 100\%$ critical value changes with both k and λ, for fixed n. If λ is outside the range $[1/3, 3/2]$, the chi-squared significance level becomes an increasingly more conservative approximation to the exact level as k grows. Also as λ moves further away from $[1/3, 3/2]$, the rate of increase (with k) of this error grows.

Case 1: $\lambda < 0$

To explain why this occurs, we consider first a fixed $\lambda < 0$. Clearly as k increases for fixed n, the proportion of possible cells containing small fre-

Table 6.2. Percentage of Possible Partitions of n Counts into k Cells Containing at Least One Cell with a Frequency of 0 or 1 (percentages rounded to nearest whole number)

	k			
n	2	3	4	5
10	36%	77%	97%	100%
20	19%	48%	74%	91%

Source: Read (1984b).

quencies will increase. Table 6.2 gives the percentage of possible partitions $\mathbf{x} = (x_1, x_2, \ldots, x_k)$ of n counts into k cells containing at least one cell with a frequency of 0 or 1. The results of Section 6.2 (with $\delta = 0$, 1 and $m_i = n/k$; $i = 1, \ldots, k$) indicate that it is precisely these partitions that result in large values of $2I^\lambda(\mathbf{x} : \mathbf{m})$ for $\lambda < 0$. However the chi-squared critical value is fixed for all λ; therefore as k increases, a larger proportion of partitions will result in statistics with values equal to or exceeding this critical value than would occur for, say, $\lambda > 0$. The overall result will be to increase the exact level associated with the chi-squared critical value, as we observed in Section 5.3. As n increases, the percentage of this type of partition decreases (Table 6.2), as observed by the better approximations obtained in Section 5.3, when n moves from 10 to 20.

Case 2: $\lambda > 0$

For $\lambda > 0$ and moderate (say $\lambda > 1$), Section 5.3 indicates that as k increases, the chi-squared significance level will again be increasingly conservative when compared to the exact level. However the explanation is different from the case $\lambda < 0$. As k increases for fixed n, the expected cell frequencies (for the equiprobable model) $m_i = n/k$; $i = 1, \ldots, k$, decrease. Furthermore, for a fixed observed frequency $x_j = \delta$, the ratio $\delta/m_j = k\delta/n$ increases. If we let δ represent a single large observed frequency in a possible partition $\mathbf{x}$, then (for $\lambda > 0$) the minimum asymptotic value of the statistic $2I^\lambda(\mathbf{x} : n\mathbf{1}/k)$ increases with $k\delta/n$, and for larger λ the rate of increase is greater (Table 6.1). Therefore, since the chi-squared critical value is fixed for all λ, we observe that as k increases, a larger proportion of this type of partition will result in statistics with values equal to or exceeding the critical value.

A Generalization of the Power Results from Section 5.4

We now use the preceding discussion, regarding the functional behavior of $2I^\lambda(\mathbf{x} : \mathbf{m})$ for a single large or a single small observed frequency, to explain further the exact power results of Section 5.4. Recall that in Section 5.4 we computed the exact power of the test statistics for the bump or dip alternative model (5.16), where one of the null probabilities (from the equiprobable null model) is perturbed and the others are adjusted appropriately to ensure that they still sum to 1. These exact power computations indicated that for alternatives with a single large (small) expected cell frequency, the power of the test based on $2I^{\lambda}(\mathbf{x}:n\mathbf{1}/k)$, with exact significance level $\alpha 100\%$, increases as λ increases (decreases). This coincides with the conclusions discussed earlier regarding the type of partition associated with values of the statistic that will equal or exceed the test critical value for different λ: For $\lambda \gg 0$ ($\lambda \ll 0$), the large statistic values will be associated mainly with partitions containing single large (small) observed frequencies. Therefore by looking at the traditional goodness-of-fit statistics of Section 2.2 embedded in the power-divergence family, we are able to determine when and why some statistics are more powerful than others.

When we move away from the equiprobable null hypothesis, there are a myriad of potential choices for null and alternative models. In this general setting, it is not only difficult to make global recommendations, but there is also a substantial increase in calculations required to compute exact significance levels. As a result very few articles appear in the literature with recommendations regarding which test statistic should be used in such situations, or when to use the asymptotic distributional approximations (Ivchenko and Medvedev, 1978).

However, the arguments of this chapter still hold for the general hypothesis where the null probabilities are not all equal, by considering partitions $\mathbf{x} = (x_1, x_2, \ldots, x_k)$ of n with a single large (or small) frequency relative to the expected frequency (Section 6.2). Therefore we expect that in small samples, the exact power of the test based on $2I^{\lambda}(\mathbf{X}:\mathbf{m})$ increases as λ increases (decreases) for alternatives with a single large (small) ratio of the alternative/null expected cell frequencies while the other ratios are all near 1. This conclusion is in agreement with the power approximations of Drost et al. (1987). For $|\lambda|$ large, we also expect that generally the chi-squared critical values will be conservative for the same reasons as described in the equiprobable case.

Some specific examples of null models that are not equiprobable are discussed in the next section, together with the consequences this has for testing the fit of these models.

6.4. Three Illustrations

Example 1: Dumping-Syndrome Data

In light of the results of Section 6.3, we now return to the dumping-syndrome data in Table 3.4. Table 3.6 contains the estimated expected cell frequencies for three minimum distance estimates based on the model of homogeneity of proportions. For illustrative purposes we shall consider the power-divergence statistic with expected cell frequencies based on maximum likelihood estimates (MLEs); the results for the other minimum distance estimates given by Table 3.6 are qualitatively similar for this example.

Table 3.5 contains the values of the power-divergence goodness-of-fit statistic for $\lambda \in [-5, 5]$ based on the MLEs. We see that as λ increases from -5 to 5, the value of $2I^{\lambda}(\mathbf{x}:\mathbf{m})$ decreases. This phenomenon can be explained by examining the ratios of the observed to expected frequencies in each cell of the contingency table. Define

$$g(x_i, m_i) = \text{sgn}(x_i - m_i)\max(x_i/m_i, m_i/x_i); \qquad i = 1, \ldots, k, \qquad (6.10)$$

then $|g(x_i, m_i)|$ represents the maximum of x_i/m_i and m_i/x_i for $i = 1, \ldots, k$; the associated sign in (6.10) indicates which of these two ratios is providing the maximum. Table 6.3 gives each value of $g(x_i, \hat{m}_i)$ based on the observed frequency x_i from Table 3.4, and the MLE of the expected frequency $\hat{m}_i$ given by the entry associated with $\lambda = 0$ in Table 3.6.

The two largest values of $|g(x_i, \hat{m}_i)|$ occur for operation A_1 with moderate dumping severity and for operation A_2 with slight dumping severity. Both these values are due to large ratios of $\hat{m}_i/x_i$; consequently, from the results of Section 6.3, these two values will tend to dominate $2I^{\lambda}(\mathbf{x}:\hat{\mathbf{m}})$ as λ becomes large and negative. This can be observed from Table 3.5 by comparing the magnitudes of the values $2I^{-\lambda}(\mathbf{x}:\hat{\mathbf{m}})$ to $2I^{\lambda}(\mathbf{x}:\hat{\mathbf{m}})$, for $\lambda \in [0, 5]$. Using the results of Chapters 4 and 5, we conclude this example by recommending that without further a priori information regarding important alternative models involving these highly influential cells, we should base our test statistic on $\lambda = 2/3$. Using a chi-squared critical value of $\chi^2_6(0.05) = 12.59$, we accept the null model, which proposes homogeneity of dumping-severity proportions for each operation.

In general the relative magnitudes of $2I^{\lambda}(\mathbf{x}:\mathbf{m})$ for $\lambda \in [-5, 5]$ indicate whether there are any particularly large ratios of x_i/m_i or m_i/x_i. If the growth of $2I^{\lambda}(\mathbf{x}:\mathbf{m})$ with $|\lambda|$ is faster for $\lambda < 0$ than for $\lambda > 0$ (as in Table 3.5), then the cells are likely dominated by a large ratio of m_i/x_i (namely, 1.58 from Table

Table 6.3. Values of $g(x_i, \hat{m}_i)$ Defined in (6.10) for the Dumping-Syndrome Data with x_i from Table 3.4 and $\hat{m}_i$ from Table 3.6 Associated with $\lambda = 0$

	Dumping severity		
Operation	None	Slight	Moderate
A_1	1.10	−1.06	−1.58
A_2	1.14	−1.40	1.09
A_3	−1.09	1.18	−1.06
A_4	1.16	1.15	1.30

6.3). Conversely, if the growth of $2I^\lambda(\mathbf{x}:\mathbf{m})$ with $|\lambda|$ is faster for $\lambda > 0$ than for $\lambda < 0$, then the cells are likely dominated by a large ratio of x_i/m_i. Finally, if there is little change in the values of $2I^\lambda(\mathbf{x}:\mathbf{m})$ for $\lambda \in [-5, 5]$, then the magnitude of $g(x_i, m_i)$ is probably fairly similar across all cells.

Example 2: Homicide Data

We illustrate this last case with an example from Haberman (1978, pp. 82–83), in which he considers the distribution of homicides in the United States during 1970, and proposes the loglinear model

$$H_0\text{: } \log(m_i/z_i) = \alpha + \beta i; \qquad i = 1, \ldots, 12. \tag{6.11}$$

Here z_i represents the number of days in the month i, and the monthly frequencies X_i are assumed independent Poisson random variables with means m_i; $i = 1, \ldots, 12$. Table 6.4 gives each observed monthly frequency x_i together with the MLE $\hat{m}_i$ under H_0, and the value $g(x_i, \hat{m}_i)$ from (6.10). In this case we see that every value $|g(x_i, \hat{m}_i)|$ is very close to 1. This indicates that every ratio $x_i/\hat{m}_i$ (or $\hat{m}_i/x_i$) is near 1, and there should be no great change in the value of the power-divergence statistic as λ varies, since no single ratio of observed to expected frequencies dominates. This is substantiated by the values of $2I^\lambda(\mathbf{x}:\hat{\mathbf{m}})$ in Table 6.5, which are all approximately 18. Consequently our sensitivity analysis tells us that it makes very little difference which statistic is chosen here. With 10% and 5% critical values of $\chi^2_{10}(0.10) = 16.0$ and

Table 6.4. Monthly Distribution of Homicides in the United States in 1970

Month	Observed (x_i)	Expected ($\hat{m}_i$)	$g(x_i, \hat{m}_i)$
January	1,318	1,323.2	−1.00
February	1,229	1,211.9	1.01
March	1,327	1,360.6	−1.03
April	1,257	1,335.2	−1.06
May	1,424	1,399.0	1.02
June	1,399	1,372.9	1.02
July	1,475	1,438.6	1.03
August	1,559	1,458.8	1.07
September	1,417	1,431.5	−1.01
October	1,507	1,500.0	1.01
November	1,400	1,472.0	−1.05
December	1,534	1,542.4	−1.01

The estimated expected frequencies in column 3 and values of $g(x_i, \hat{m}_i)$ from (6.10) in column 4 are calculated assuming the loglinear model (6.11).

Source: Haberman (1978, pp. 82–83). National Center for Health Statistics (1970, pp. 1–174, 1–175).

Table 6.5. Computed Values of the Power-Divergence Statistic $2I^{\lambda}(\mathbf{x}:\hat{\mathbf{m}})$ for the Data in Table 6.4

				λ					
-5	-2	-1	$-1/2$	0	1/2	2/3	1	2	5
18.3	18.4	18.4	17.8	18.0	18.0	18.0	18.1	18.1	18.3

$\chi^2_{10}(0.05) = 18.3$, we would formally reject the model (6.11) (at the 10% level) regardless of our choice of λ. It is important to note however that the total number of homicides in this example is very large ($n = 16{,}846$), consequently all these test statistics have high power for detecting very small deviations from the null model (Section 4.2). Therefore even though we might formally reject the model (6.11), it may still be a useful summary of the data. The issue of compensating for large sample sizes to prevent formal rejection of adequate models is discussed further in Section 8.3.

Example 3: Memory-Recall Data

Finally we illustrate how a large value of $|g(x_i, m_i)|$ can have a substantial effect on the magnitude of change in $2I^{\lambda}(\mathbf{x}:\mathbf{m})$ for quite a small change in λ. We return to the example of Section 2.3 on the relationship between time passage and memory recall. Table 2.3 indicates that $2I^{\lambda}(\mathbf{x}:\hat{\mathbf{m}})$ is smallest for $\lambda \in [1/2, 2]$, and as λ decreases from 1/2, $2I^{\lambda}(\mathbf{x}:\hat{\mathbf{m}})$ increases rapidly. As λ increases from 2, $2I^{\lambda}(\mathbf{x}:\hat{\mathbf{m}})$ also increases, but at a much slower rate. Consequently using a 5% significance level ($\chi^2_{16}(0.05) = 26.3$) we would accept the time-trend model (2.15) based on Pearson's $X^2 = 22.7$ ($\lambda = 1$), the power-divergence statistic $2I^{2/3}(\mathbf{x}:\hat{\mathbf{m}}) = 23.1$, or the loglikelihood ratio statistic $G^2 = 24.6$ ($\lambda = 0$). However using the Neyman-modified X^2 statistic $NM^2 = 40.6$ ($\lambda = -2$) we would strongly reject the model. Looking at Table 6.6 we see the ratio $\hat{m}_i/x_i$ is very large in cells 16 and 17 where the expected frequencies are four times the observed frequencies. These two values account for the rapid increase in the magnitude of $2I^{\lambda}(\mathbf{x}:\hat{\mathbf{m}})$ as λ becomes large in the negative direction. However the largest values of $x_i/\hat{m}_i$ (in cells 12 and 13) are only about half the size of the large values of $\hat{m}_i/x_i$ observed in cells 16 and 17, and consequently we see a much slower increase in magnitude of $2I^{\lambda}(\mathbf{x}:\hat{\mathbf{m}})$ as λ becomes large in the positive direction.

To conclude this example, we note that for an a priori alternative model that proposes smaller probabilities for memory recall beyond (say) 12 months than does the loglinear time-trend model, then choosing a negative value of λ would be reasonable. However, if we look back at the observed cell frequencies of Table 2.2, we see that the influential cells 16 and 17 contain only one observation each. Consequently a slight frequency change in these cells would

Table 6.6. Values of $g(x_i, \hat{m}_i)$ Defined in (6.10) for the Memory-Recall Data of Table 2.2

Months before interview	$g(x_i, \hat{m}_i)$
1	−1.01
2	−1.27
3	1.09
4	1.44
5	−2.17
6	1.10
7	1.09
8	−2.11
9	1.03
10	1.40
11	1.07
12	1.49
13	1.98
14	−1.70
15	1.28
16	−4.32
17	−3.97
18	1.10

have a substantial effect on the values of $g(x_i, \hat{m}_i)$ for $i = 16, 17$, and hence on the values of $2I^\lambda(\mathbf{x} : \hat{\mathbf{m}})$ for negative λ. With this instability in mind, and no specific alternative proposed, we recommend using $\lambda = 2/3$ and accepting the loglinear time-trend model for memory recall given by (2.15). In Section 6.7 we discuss more general guidelines that can be applied to other examples.

6.5. Transforming for Closer Asymptotic Approximations in Contingency Tables with Some Small Expected Cell Frequencies

In Section 6.2 we observed the effect of individual cell contributions on the power-divergence statistic through calculating the minimum asymptotic value of $2I^\lambda(\mathbf{x} : \mathbf{m})$, defined by (6.4) and (6.5). Following the results of Anscombe (1981, 1985), we can obtain further insight into the effects of individual cell contributions by viewing the parameter λ as a transformation index on the individual cell frequencies. In order to facilitate distinguishing the positive contributions made by each cell to the overall statistic, Anscombe (1985) suggests redefining the power-divergence statistic (2.17) as

$$2I^{\lambda}(\mathbf{X}:\mathbf{m}) = \sum_{i=1}^{k} h^{\lambda}(X_i, m_i); \qquad -\infty < \lambda < \infty, \tag{6.12}$$

where

$$h^{\lambda}(X_i, m_i) = \frac{2}{\lambda(\lambda+1)}\left\{X_i\left[\left(\frac{X_i}{m_i}\right)^{\lambda} - 1\right] + \lambda[m_i - X_i]\right\}; \tag{6.13}$$

for $i = 1, \ldots, k$. This definition has the property that each $h^{\lambda}(X_i, m_i)$ is nonnegative and vanishes when $X_i = m_i$ (for $i = 1, \ldots, k$), however (6.12) is equivalent to (2.17) since $\sum_{i=1}^{k} X_i = \sum_{i=1}^{k} m_i$.

For $\lambda = 1$ (Pearson's X^2 statistic), (6.13) becomes

$$h^1(X_i, m_i) = (X_i - m_i)^2/m_i;$$

for $\lambda = 0$ (the loglikelihood ratio statistic G^2), (6.13) becomes

$$h^0(X_i, m_i) = 2X_i \log(X_i/m_i) + 2(m_i - X_i),$$

which is (minus twice) the Poisson loglikelihood ratio; and for $\lambda = -1/2$, (6.13) becomes

$$h^{-1/2}(X_i, m_i) = 4(\sqrt{X_i} - \sqrt{m_i})^2.$$

When there are some cells with small expected values (more details are given later), the individual cell contributions (6.13) provide a way to compare the asymptotic distributions of the power-divergence family members. For contingency tables, Anscombe (1981, p. 293) points out that "the count in a cell for which the row total and column total are both much larger than the cell expectation has approximately a Poisson distribution, and is approximately independent of the count in another such cell." Using this Poisson approximation we can determine the effect of cells with low expectations on the moments of the statistic (6.12) by studying the moments of the individual frequencies in (6.13). For example Table 6.7 gives the mean, variance, and third central moment of (6.13) for $\lambda = 0$, $\lambda = 2/3$, and $\lambda = 1$ with cell expectations between 0.1 and 10.0.

Anscombe (1981) notes that based on these individual transformed Poisson-cell moments (Table 6.7), we can infer (approximately) the first three moments of the associated goodness-of-fit statistics when a few cells have small expectations (while the rest of the expectations and the marginal totals of the contingency table are all large). This is done by taking the usual asymptotic chi-squared moments for the table and adding, for each cell with low expectation, the difference between the moments given in Table 6.7 and the limiting values of these moments (for large m_i) in the last column.

From Table 6.7 we see that for cell expectations greater than 1 the transformation $\lambda = 2/3$ provides moments more closely approximating the limiting values than either $\lambda = 0$ or $\lambda = 1$. This indicates that among these three cases, the power-divergence statistic (6.12) with $\lambda = 2/3$ has moments closest to the asymptotic chi-squared moments. Apart from this one special case, the effect of power transformations on the distribution of individual cell frequencies

Table 6.7. Moments for the Power Transformation (6.13) of a Single Poisson-Distributed Cell

	Cell expectation							
	0.1	0.2	0.5	1	2	5	10	∞
				$\lambda = 0$				
mean	0.47	0.70	1.01	1.15	1.14	1.05	1.02	1
var	0.86	0.73	0.73	1.36	2.23	2.27	2.09	2
μ_3	3.66	3.36	3.00	2.66	5.81	9.88	8.64	8
				$\lambda = 2/3$				
mean	0.70	0.81	0.93	0.98	1.00	1.00	1.00	1
var	4.46	3.11	2.18	1.99	2.00	2.01	2.01	2
μ_3	58.62	35.61	20.12	13.76	10.59	9.06	8.53	8
				$\lambda = 1$				
mean	1.00	1.00	1.00	1.00	1.00	1.00	1.00	1
var	12.00	7.00	4.00	3.00	2.50	2.20	2.10	2
μ_3	328.00	143.00	56.00	31.00	19.25	12.44	10.21	8

Source: Anscombe (1981, p. 294; 1985).

deserves further research, with $\lambda = 2/3$ appearing to offer a promising transformation (see Section 8.2 for related work).

6.6. A Geometric Interpretation of the Power-Divergence Statistic

It has been shown in the previous section that the power-divergence statistic (2.17) can be written as

$$2I^{\lambda}(\mathbf{X}:\mathbf{m}) = \sum_{i=1}^{k} h^{\lambda}(X_i, m_i),$$

where $h^{\lambda}(X_i, m_i)$ is given by (6.13) to be

$$h^{\lambda}(X_i, m_i) = \frac{2}{\lambda(\lambda+1)}\left\{X_i\left[\left(\frac{X_i}{m_i}\right)^{\lambda} - 1\right] + \lambda[m_i - X_i]\right\}.$$

Now consider the related function

$$h(y) = \frac{2}{\lambda(\lambda+1)}[y^{\lambda+1} - 1 - (\lambda+1)(y-1)]. \qquad (6.14)$$

Then $h'(y) = 2(y^{\lambda} - 1)/\lambda$ and $h''(y) = 2y^{\lambda-1}$, which shows that for $y \geq 0$, $h(y)$ is a convex function that achieves a minimum at $y = 1$. In fact, $h(1) = 0$, proving that for $y \geq 0$, $h(y) \geq 0$ (and since $h^{\lambda}(X_i, m_i) = m_i h(X_i/m_i)$, that $h^{\lambda} \geq 0$).

The Taylor-series expansion of $h(y)$ about $y = 1$ is given by

$$h(y) \simeq h(1) + (y-1)h'(1) + (y-1)^2 h''(1)/2! = (y-1)^2, \tag{6.15}$$

and hence

$$\begin{aligned} h^\lambda(X_i, m_i) &= m_i h(X_i/m_i) \\ &\simeq (X_i - m_i)^2/m_i \\ &= h^1(X_i, m_i). \end{aligned}$$

Therefore this approximate representation shows that algebraically, as well as statistically, all members of the power-divergence family behave like Pearson's X^2 (this approximation was used extensively in Chapter 4). Geometrically, Pearson's $X^2 = \sum_{i=1}^k (X_i - m_i)^2/m_i$ is a weighted sum of squared differences. When the X_is are independent Poisson random variables with the mean of X_i being m_i; $i = 1, \ldots, k$, then the weighting is inversely proportional to the variance.

In fact, these results are an oversimplification, caused by the inadequacy of the Taylor-series polynomial approximation to fractional powers. To see this, define

$$a(z) = \frac{2}{\lambda(\lambda+1)}[z^{(\lambda+1)/\lambda} - 1 - (\lambda+1)(z^{1/\lambda} - 1)].$$

Then from (6.14), $h(y) = a(y^\lambda)$, and the Taylor-series expansion of $a(z)$ about $z = 1$ is

$$a(z) \simeq a(1) + (z-1)a'(1) + (z-1)^2 a''(1)/2! = [(z-1)/\lambda]^2;$$

therefore

$$h(y) \simeq \left[\frac{y^\lambda - 1}{\lambda}\right]^2. \tag{6.16}$$

Using approximation (6.16) for (6.14), (6.13) becomes

$$h^\lambda(X_i, m_i) \simeq m_i \left[\frac{(X_i/m_i)^\lambda - 1}{\lambda}\right]^2,$$

and hence the power-divergence statistic becomes

$$2I^\lambda(\mathbf{X} : \mathbf{m}) \simeq \sum_{i=1}^k \frac{(X_i^\lambda - m_i^\lambda)^2}{\lambda^2 m_i^{2\lambda-1}}. \tag{6.17}$$

Therefore the power-divergence statistic can be interpreted geometrically as (roughly speaking) a sum of weighted squared differences, between the observed frequencies raised to a power and the expected frequencies raised to the same power.

Approximation (6.17) provides a partial explanation as to why the power-divergence statistic with $\lambda = 2/3$, approximated as

$$2I^{2/3}(\mathbf{X}:\mathbf{m}) \simeq \sum_{i=1}^{k} \frac{(X_i^{2/3} - m_i^{2/3})^2}{(4/9)m_i^{1/3}}, \tag{6.18}$$

performs so well in the previous chapters of this book. Anscombe (1953) points out that the 2/3 power of a Poisson random variable has asymptotically zero third central moment, so that each term on the right-hand side of (6.18) is very much like a chi-squared random variable with one degree of freedom. Anscombe (1985) modifies (6.18) slightly to

$$\sum_{i=1}^{k} \frac{[(X_i + 1/4)^{2/3} - (m_i + 1/12)^{2/3})]^2}{(4/9)m_i^{1/3}},$$

so that it behaves more like a chi-squared random variable when m_i is small.

6.7. Which Test Statistic?

In this chapter we have seen that choosing λ to test H_0 depends on the type of departure from H_0 we wish to measure. Departures involving large ratios of the alternative to null expected frequencies in one or two cells are best detected using large values of λ, say $\lambda = 5$. Conversely, departures involving ratios of alternative to null expected frequencies that are close to 0 in one or two cells are best detected using large negative values of λ, say $\lambda = -5$. In general, we do not recommend using statistics for which $|\lambda|$ is larger than 5, since they are extremely sensitive to the largest (smallest) ratio of observed to expected frequencies, and essentially ignore the contributions of all other cells. This is illustrated in the extreme cases by equations (6.8) and (6.9).

Using the continuous parameter λ, it is possible to trade off the effects noted in Section 6.2 for X^2, G^2, and F^2, by using tables like Table 6.1 to choose a statistic according to the emphasis we wish to attach to large or small ratios of x_i/m_i. However, we emphasize that the choice of an alternative model (and hence of λ) should be made before looking at the data. If λ is chosen to highlight some characteristic after observing the sample, then any significance test will not have the specified size. Instead of a random sample from the possible realizations of the power-divergence statistic, we shall have effectively biased the sampling in favor of large values of the statistic.

The question of how to determine the appropriate critical value for the test statistic must not be overlooked when choosing λ. Section 5.5 provides recommendations for finite samples, and Sections 6.5 and 6.6 (see in particular (6.18)) explain why the power-divergence statistic with $\lambda = 2/3$ will be well approximated by the chi-squared distribution in many situations.

From the data-analytic point of view, comparing the computed values of the power-divergence statistic for different values of λ (e.g., Tables 3.5, 6.5, and 2.3) can help to assess the extent of departure from the model. For example, a table of values for $2I^{\lambda}(\mathbf{x}:\mathbf{m})$ that are very close together over the range

$\lambda \in [-5, 5]$ (e.g., Table 6.5) implies no single cell is making a major contribution to the lack of fit, and therefore the choice of statistic to test the null model is not critical. However, if there is a wide discrepancy between the values for $2I^{\lambda}(\mathbf{x}:\mathbf{m})$ (e.g., Table 2.3), then we conclude that one or two cells have large relative departures from the expected cell frequencies. But we should not then perform a formal significance test based on a value of λ chosen from such a table, since this procedure will produce meaningless significance levels as discussed earlier.

If we do not wish to place too much weight on any single cell departure, but wish to look more for overall lack of fit, then choosing $|\lambda|$ small is advisable. However $\lambda = 0$ is not necessarily the optimum choice in this setting; for example, Section 5.4 indicates that choosing $\lambda \in [1/3, 2/3]$ gives most reasonable power against both bump and dip alternatives (which are special cases of a large single-cell departure from the equiprobable null model). Therefore we recommend the use of $\lambda = 2/3$

$$2I^{2/3}(\mathbf{x}:\mathbf{m}) = \frac{9}{5}\sum_{i=1}^{k} x_i\left[\left(\frac{x_i}{m_i}\right)^{2/3} - 1\right]$$

as a good compromise that will give some protection against both types of extreme departures discussed earlier. This choice is consistent with the recommendations of Section 4.5 (regarding performance in large samples) and Section 5.5 (regarding the ease of calculating the critical value for the test statistic, provided the minimum expected cell frequency is greater than 1).

CHAPTER 7

Links with Other Test Statistics and Measures of Divergence

In the preceding chapters, we have concentrated on test statistics that measure the degree of divergence between the observed frequencies for a group of cells and the corresponding expected frequencies based on some null model. Throughout we have assumed that we are dealing with discrete counts. If the variable under observation has a continuous (in general, multivariate) distribution, we assume that the outcomes have been grouped into mutually exclusive intervals (in general, compact regions) or cells, whose union contains the support (i.e., the region where the density is positive) of the random variable. We can then use the frequencies with which observations from a sample fall in these cells to perform a goodness-of-fit test (as described in the previous chapters). Such grouping of continuous data necessarily results in some loss of information, and we may wish to use an alternative test statistic that is more efficient in these circumstances. The first two sections of this chapter are devoted to discussing test statistics that are designed for continuous data. In Section 7.3 we draw some comparisons between these statistics and the power-divergence statistic based on the grouping described earlier.

In Section 7.4 the focus moves from goodness-of-fit test statistics for continuous data to the closely related literature on measures of diversity (or entropy) and directed divergence for discrete distributions. While goodness-of-fit statistics compare an empirical distribution with a hypothetical probability distribution, measures of directed divergence compare the closeness of two (or more) discrete theoretical probability distributions. In this section we link the study and characterization of these measures to the power-divergence statistic.

7.1. Test Statistics Based on Quantiles and Spacings

Throughout this section we shall assume that $Y_1, Y_2, \ldots, Y_n$ are independent and identically distributed random variables with a continuous distribution function $F(y)$, and support $(-\infty, \infty)$. Denote the order statistics $Y_{(1)} \leq Y_{(2)} \leq \cdots \leq Y_{(n)}$ and define $Y_{(0)} = -\infty$, $Y_{(n+1)} = \infty$. In general we wish to test the null hypothesis

$$H_0\colon F(y) = F_0(y; \boldsymbol{\theta}), \tag{7.1}$$

where $F_0(y; \boldsymbol{\theta})$ is a specified distribution function and $\boldsymbol{\theta}$ is a vector of parameters (which may be unknown and require estimation).

In order to use the power-divergence statistic for testing hypothesis (7.1), it is necessary to group the continuous data according to a set of $k + 1$ boundaries

$$-\infty \leq b_1 < b_2 < \cdots < b_{k+1} \leq \infty,$$

where $F_0(b_1; \boldsymbol{\theta}^*) = 0$ and $F_0(b_{k+1}; \boldsymbol{\theta}^*) = 1$ and $\boldsymbol{\theta}^*$ is the true value of $\boldsymbol{\theta}$. Thus there are k mutually exclusive and exhaustive cells $\{(b_i, b_{i+1}]; i = 1, \ldots, k\}$ and we define

$$X_i = \# \; Y_j\text{s} \in (b_i, b_{i+1}]; \qquad j = 1, \ldots, n, \quad i = 1, \ldots, k$$

and

$$\pi_{0i}(\boldsymbol{\theta}) = F_0(b_{i+1}; \boldsymbol{\theta}) - F_0(b_i; \boldsymbol{\theta}); \qquad i = 1, \ldots, k,$$

to obtain the power-divergence statistic (2.16):

$$2nI^{\lambda}(\mathbf{X}/n : \boldsymbol{\pi}_0(\hat{\boldsymbol{\theta}})) = \frac{2}{\lambda(\lambda+1)} \sum_{i=1}^{k} X_i \left[\left(\frac{X_i}{n\pi_{0i}(\hat{\boldsymbol{\theta}})} \right)^{\lambda} - 1 \right]; \quad -\infty < \lambda < \infty. \tag{7.2}$$

Here $\hat{\boldsymbol{\theta}}$ is some BAN estimator of $\boldsymbol{\theta}$, based on the grouped data. Little has been said yet about the choice of boundaries $b_1, b_2, \ldots, b_{k+1}$; it is generally agreed that they should be chosen so that the resulting cells are equiprobable under the null hypothesis (Sections 3 and 4 of the Historical Perspective).

There are three main alternative types of test statistic that may be used in place of the power-divergence statistic and do not rely on (a potentially arbitrary) grouping of the observations. These test statistics are based on:

(a) The order statistics or sample quantiles

$$Y_{(i)}; \qquad i = 0, \ldots, n+1 \tag{7.3}$$

(e.g., Bofinger, 1973; Miyamoto, 1976).

(b) The transformed spacings (or generalizations thereof)

$$V_i = F_0(Y_{(i)}; \boldsymbol{\theta}) - F_0(Y_{(i-1)}; \boldsymbol{\theta}); \qquad i = 1, \ldots, n+1 \tag{7.4}$$

(e.g., Greenwood, 1946; Pyke, 1965; Cressie, 1976; Hall, 1986).

(c) The empirical distribution function (EDF)

$$F_n(y) = (\# \, Y_i\text{s} \le y)/n \tag{7.5}$$

(e.g., the Kolmogorov-Smirnov statistic and Cramér-von Mises statistic reviewed by Stephens, 1986a).

Gebert and Kale (1969) indicate that test statistics of type (b) are useful for detecting departures from hypothesized density functions (such as $f_0(y;\boldsymbol{\theta}) = dF_0(y;\boldsymbol{\theta})/dy$) while test statistics of type (c) are useful for detecting departures from hypothesized distribution functions (such as $F_0(y;\boldsymbol{\theta})$). We now consider these three types of test statistics in more detail.

Test Statistics Based on Sample Quantiles

First we consider a test statistic based on a subset of the order statistics $Y_{(1)} \le Y_{(2)} \le \cdots \le Y_{(n)}$. If m_i is the integer part of $(n+1)\gamma_i$; $i = 1, \ldots, k+1$ for some fixed k and γ_i $(0 \equiv \gamma_1 < \cdots < \gamma_{k+1} \equiv 1)$, Bofinger (1973) defines the sample quantile statistic

$$X_Q^2 = n \sum_{i=1}^{k} (F_0(Y_{(m_{i+1})};\bar{\boldsymbol{\theta}}) - F_0(Y_{(m_i)};\bar{\boldsymbol{\theta}}) - \pi_{0i})^2/\pi_{0i}, \tag{7.6}$$

where $\pi_{0i} \equiv \gamma_{i+1} - \gamma_i$; $i = 1, \ldots, k$, $F_0(Y_{(0)};\bar{\boldsymbol{\theta}}) \equiv 0$, $F_0(Y_{(n+1)};\bar{\boldsymbol{\theta}}) \equiv 1$, and $\bar{\boldsymbol{\theta}}$ is an estimator of $\boldsymbol{\theta}$. (Bofinger's definition of m_i differs slightly from ours.)

The similarity of (7.6) and Pearson's X^2 statistic for grouped data (i.e., (7.2) with $\lambda = 1$) can be observed when b_i from (7.2) is defined to satisfy $F_0(b_i;\boldsymbol{\theta}^*) = \gamma_i$; $i = 1, \ldots, k+1$. Provided $F_0(y;\boldsymbol{\theta}^*)$ is continuous and strictly increasing in y in the neighborhood of b_i, these b_i will be unique, although not computable in practice since $\boldsymbol{\theta}^*$, the true value of $\boldsymbol{\theta}$, is unknown. If, in addition, certain standard regularity conditions are required of $\bar{\boldsymbol{\theta}}$ and F_0, then Bofinger (1973) shows that Pearson's X^2 and the sample quantile statistic X_Q^2 of (7.6) are asymptotically equivalent both under the null hypothesis and under certain local alternatives (for further details see Miyamoto, 1976; Durbin, 1978).

A consequence of this asymptotic equivalence is that a preference for using either a sample quantile statistic or power-divergence statistic (7.2) will have to be based on criteria that address more "distant" alternatives and small-sample considerations. Bofinger (1973) suggests a practical advantage is that (7.6) avoids the difficulties of choosing the boundaries b_i; $i = 1, \ldots, k+1$. However this advantage appears unfounded since it is still necessary to choose the γ_is. Furthermore the arbitrariness of the b_is is removed if they are chosen to give equal cell probabilities $\pi_{0i}(\boldsymbol{\theta}) = 1/k$; $i = 1, \ldots, k$, a choice frequently recommended in the literature (Section 3 of the Historical Perspective; Section 4 also provides a discussion of how to estimate $\boldsymbol{\theta}$ and choose data-dependent cell boundaries when $\boldsymbol{\theta}$ is unknown).

In conclusion the sample quantile approach has no obvious advantages over the power-divergence statistic (7.2) based on grouped data.

Test Statistics Based on First-Order Spacings

In the following discussion of spacings, we shall assume that the hypothesis (7.1) is simple (i.e., $\boldsymbol{\theta}$ is completely specified within $F_0(y)$). Applying the probability integral transformation $F_0(y)$ to the sample values $Y_1, Y_2, \ldots, Y_n$ allows us to assume without loss of generality that the null hypothesis (7.1) specifies the uniform distribution $U(0, 1)$ on the unit interval. In other words we wish to test

$$H_0: G = U(0, 1), \tag{7.7}$$

where $G(u)$; $0 \le u \le 1$ is the distribution function of $U_i = F_0(Y_i)$; $i = 1, \ldots, n$.

If $U_{(1)} \le U_{(2)} \le \cdots \le U_{(n)}$ are the order statistics from the transformed sample, then the (first-order) spacings are defined by (7.4) to be $V_i = U_{(i)} - U_{(i-1)}$; $i = 1, \ldots, n + 1$ where $U_{(0)} \equiv 0$ and $U_{(n+1)} \equiv 1$. To test (7.7) we can define an analogous statistic to (7.2) using spacings by replacing the observed proportions X_i/n by the spacings V_i and the expected probabilities π_{0i} by the expectations $E(V_i) = 1/(n + 1)$; $i = 1, \ldots, n + 1$. Hence there is a version of the power-divergence statistic (7.2) for spacings:

$$2nI^\lambda(\mathbf{V} : E(\mathbf{V})) = \frac{2n}{\lambda(\lambda + 1)} \sum_{i=1}^{n+1} V_i[((n + 1)V_i)^\lambda - 1]; \qquad -\infty < \lambda < \infty. \tag{7.8}$$

Test statistics equivalent to (7.8) (up to a multiplicative and an additive constant) have been considered from various perspectives by many authors (e.g., Greenwood, 1946, for $\lambda = 1$; Kimball, 1947, and Darling, 1953, for $\lambda \ge -1$, $\lambda \ne 0$; Kale and Godambe, 1967, for $\lambda = 0$ and $\lambda = -1$; Kirmani, 1973, for $\lambda = -1/2$; Kirmani and Alam, 1974, for $\lambda > -1$). The power-divergence statistic for spacings (7.8) has been shown to be asymptotically normally distributed for $\lambda \ge -1$ (Darling, 1953, for $\lambda \ge -1$, $\lambda \ne 0$; Gebert and Kale, 1969, for $\lambda = 0$), but not in general for $\lambda < -1$ (e.g., Darling, 1953, shows that for $\lambda = -2$ a quasi-stable law of exponent 1 is obtained). These distributional results correspond to the asymptotic theory under the sparseness assumptions described in Section 4.3.

Parallels with the Results of Chapter 4

There are many parallels between the asymptotic theory of the statistic (7.8) and the results under the sparseness assumptions described in Section 4.3. For example, Holst (1972) recognizes that a generalization of his limit theorem (Section 4.3) could be obtained using analogous methods to those described by Pyke (1965) for spacings (Holst and Rao, 1981, provide a more recent discussion of this point). The parallels between the results for sample quantiles versus spacings on the one hand, and multinomials with fixed (Section 4.1) versus increasing (Section 4.3) numbers of cells on the other, become clear once we notice that the sample quantile statistics involve just a subset of the spacings. For the sample quantile statistic X_Q^2 we can set the number of cells $k = n + 1$ and $\gamma_i = (i - 1)/(n + 1)$ giving $\pi_{0i} = \gamma_{i+1} - \gamma_i = 1/(n + 1)$; $i = 1, \ldots, n + 1$, which makes (7.6) equivalent to the power-divergence statistic for spacings (7.8) with $\lambda = 1$ (assuming $\boldsymbol{\theta}$ is fully specified under H_0). Furthermore, as the number of sample quantiles used in calculating (7.6) increases from $k + 1$ (k assumed fixed as $n \to \infty$) to $n + 2$, the asymptotic distribution of X_Q^2 changes from chi-squared (Bofinger, 1973) to normal (Darling, 1953). This progression from chi-squared to normal parallels the discrete multinomial results of Chapter 4 where the asymptotic distribution of Pearson's X^2 is chi-squared (and equivalent to X_Q^2) for the classical (fixed-cells) assumptions but changes to the normal distribution for the sparseness assumptions.

In Section 4.3 we saw that Pearson's X^2 ($\lambda = 1$) is maximally efficient among the power-divergence family for testing local alternatives of order $O(n^{-1/4})$ against the equiprobable null model (hypothesis (4.17)). From the parallels with spacings described earlier, it should be no surprise that Gebert (1968) and Kirmani and Alam (1974) show Greenwood's statistic (i.e., (7.8) with $\lambda = 1$ up to a constant) has maximum asymptotic power among the power-divergence family (7.8) for any $\lambda \geq -1$ using similar local alternatives defined on the spacings.

Jammalamadaka and Tiwari (1985) compare Greenwood's statistic to Pearson's X^2 with $k = n$ cells, and show that in this case X^2 is only 25% as efficient as Greenwood's statistic for the same local alternatives discussed earlier. It is interesting to note that X^2 compares the difference of the observed and expected cell frequencies, holding the expected frequencies equal to 1; while Greenwood's statistic compares the difference of the observed and expected cell lengths, holding the observed number in each cell equal to 1.

mth-Order Spacings and Overlapping Cells

Several authors have proposed generalizations of first-order uniform spacings to mth-order uniform spacings (for references see Hall, 1986; Stephens, 1986b). Cressie (1976) considers test statistics of the form

$$S_n^{(m)} = \sum_{i=1}^{n+2-m} g((n + 1)V_i^{(m)}); \qquad m \leq n + 1 \tag{7.9}$$

where g is a "smooth" function, and

$$V_i^{(m)} = U_{(i-1+m)} - U_{(i-1)}; \qquad i = 1, \ldots, n + 2 - m$$

are the (overlapping) mth-order spacings or mth-order gaps. (Disjoint higher-order spacings are considered by del Pino, 1979, but they result in less powerful test statistics than (7.9); see Cressie, 1979; Hall, 1986.)

Asymptotic normality of $S_n^{(m)}$ is proved for fixed m, and for $m = o(n^{1/3})$ when $g(x) = \log(x)$, by Cressie (1976). The general asymptotic ($n \to \infty$) result for any

m, which involves a mixture of normal and chi-squared distributions, is established by Hall (1986) and Guttorp and Lockhart (1988).

In determining which m to use, the principle conclusion is that larger values of m result in (asymptotically) more powerful tests (Dudewicz and van der Meulen, 1981; Hall, 1986). A similar result for Pearson's X^2 is discussed by Hall (1985); he shows that a modification allowing overlapping cells produces a test statistic with superior power to the statistic based on the standard disjoint cells. Hall (1985) gives further conditions under which this modified statistic is more powerful than the Cramér-von Mises statistic discussed in Section 7.2.

The optimality of the Greenwood statistic (i.e., (7.8) with $\lambda = 1$) among members of the power-divergence family (7.8) for first-order spacings ($m = 1$) carries over to the mth-order spacings. This is immediate from the stronger result in Cressie (1979); he shows that $g(x) = x^2$ in (7.9) is optimal (in terms of Pitman asymptotic relative efficiency). This work is unreferenced by Rao and Kuo (1984), who derive the same result for alternatives more general than the step functions considered by Cressie (1979). In the case of nonoverlapping spacings, Jammalamadaka and Tiwari (1987) generalize their work for first-order spacings to show that Pearson's X^2 based on k cells (with $n/k \to m$ as $n \to \infty$ and $k \to \infty$) is $[m^2/(m+1)^2]100\%$ as efficient as Greenwood's statistic based on nonoverlapping mth-order spacings. A detailed discussion of how m should increase with the sample size n is provided by Hall (1986) but is beyond the scope of this chapter.

In conclusion, test statistics based on first-order spacings have asymptotic power that is comparable to the power-divergence statistic (7.2) with $k = n$ cells (e.g., Hall, 1985). The results for mth-order spacings indicate that a modification to (7.2) allowing overlapping cells should produce a power-divergence statistic with power comparable to test statistics based on overlapping mth-order spacings. More research is needed before further recommendations can be made on overlapping cells.

7.2. A Continuous Analogue to the Discrete Test Statistic

In this section we develop a continuous analogue to the "discrete" power-divergence test statistic (7.2), which aids in understanding how these test statistics differ in their general form from statistics based on the EDF (7.5).

Using the same notation as in Section 7.1, we define the empirical and hypothesized grouped distribution functions to be

$$\tilde{F}_n(y) = \frac{1}{n}\sum_{j=1}^{i} X_j \qquad \text{for } b_i \le y < b_{i+1}, \quad i = 1, \ldots, k \tag{7.10}$$

and

$$\tilde{F}_0(y) = \sum_{j=1}^{i} \pi_{0j} \qquad \text{for } b_i \le y < b_{i+1}, \quad i = 1, \ldots, k, \tag{7.11}$$

respectively, where $X_1, X_2, \ldots, X_k$ are the discrete cell frequencies based on the continuous random variables $Y_1, Y_2, \ldots, Y_n$. Similarly define

$$\tilde{f}_n(y) = (\tilde{F}_n(y + h/2) - \tilde{F}_n(y - h/2))/h$$

and

$$\tilde{f}_0(y) = (\tilde{F}_0(y + h/2) - \tilde{F}_0(y - h/2))/h$$

to be the empirical and hypothesized grouped density functions. Provided h satisfies the inequality

$$0 < h/2 < \min_{1 \le i \le k} (b_{i+1} - b_i),$$

$X_i = nh\tilde{f}_n(b_i)$ and $n\pi_{0i} = nh\tilde{f}_0(b_i)$ for $i = 1, \ldots, k$. Consequently (7.2) can be written as

$$2nI^{\lambda}(\tilde{f}_n : \tilde{f}_0) = \frac{2n}{\lambda(\lambda + 1)} \int_{-\infty}^{\infty} \left[\left(\frac{\tilde{f}_n(y)}{\tilde{f}_0(y)} \right)^{\lambda} - 1 \right] d\tilde{F}_n(y). \tag{7.12}$$

Substituting the usual empirical and hypothesized distribution functions for $Y_1, Y_2, \ldots, Y_n$ (i.e., $F_n(y)$ defined by (7.5) and $F_0(y)$ defined by (7.1)) we obtain a continuous analogue to (7.12), i.e.,

$$2nI^{\lambda}(f_n : f_0) = \frac{2n}{\lambda(\lambda + 1)} \int_{-\infty}^{\infty} f_n(y) \left[\left(\frac{f_n(y)}{f_0(y)} \right)^{\lambda} - 1 \right] dy \tag{7.13}$$

where

$$f_n(y) = (F_n(y + h/2) - F_n(y - h/2))/h \tag{7.14}$$

is the naive "histogram" kernel density estimator and

$$f_0(y) = dF_0(y)/dy.$$

As n and k increase, the "discrete" statistic (7.12) measures the ratio of arbitrarily small increments of the empirical and hypothesized distribution functions. In the limit, these increments will be the hypothesized and empirical density functions. Recalling the results under the sparseness assumptions of Section 4.3 (where k becomes large with n) it should come as no surprise that Bickel and Rosenblatt (1973) show for $\lambda = 1$ that the "continuous" power-divergence statistic (7.13) attains an asymptotic normal distribution assuming certain conditions, although the rate of convergence is slow (Lewis et al. 1977). Bickel and Rosenblatt use a more general kernel density estimator f_n in place of the naive histogram estimator given by (7.14); it is an open problem to generalize their results to the power-divergence test statistic (7.13) for other values of λ.

Another special case of (7.13) is discussed by Beran (1977), who derives the asymptotic normality of the squared Hellinger distance between the hypothesized density $f_0(y)$ and some suitable density estimator $f_n(y)$. The squared Hellinger distance is a continuous version of the squared Matusita distance (Matusita, 1954), which is equivalent to (7.13) with $\lambda = -1/2$ (we have referred to the power-divergence statistic with $\lambda = -1/2$ as the Freeman-Tukey statistic throughout this book). We conjecture a similar asymptotic normal distribution will hold for general λ in (7.13).

Rather than defining continuous versions of statistics designed for discrete data, Pettitt and Stephens (1977) work in the reverse direction and define the Kolmogorov-Smirnov statistic for discrete data as

$$S = \max_{1 \le i \le k} n|\tilde{F}_n(b_i) - \tilde{F}_0(b_i)|,$$

where $\tilde{F}_n(y)$ and $\tilde{F}_0(y)$ are defined by (7.10) and (7.11), respectively. The authors provide tables comparing the power of S with Pearson's X^2, which show that, for certain "trend" alternatives where the cell probabilities increase with the cell index i, S is more powerful than X^2. However this is not the case in general for the other alternatives they consider (Stephens, 1986a, provides a more recent discussion of this statistic).

In concluding, we reiterate that the equivalence of (7.2) and (7.12) shows that the power-divergence statistic (7.2) compares changes in the hypothetical distribution with changes in the EDF. In the limit these comparisons are made between the hypothetical and empirical density functions in (7.13). Conversely the well-known EDF statistics

$$D = \sup_y |F_n(y) - F_0(y)|$$

(the Kolmogorov-Smirnov statistic), and

$$W^2 = n \int_{-\infty}^{\infty} (F_n(y) - F_0(y))^2 dF_0(y)$$

(the Cramér-von Mises statistic) compare the hypothetical distribution function directly with the EDF. While EDF statistics have been recommended on the basis of high efficiency compared to test statistics designed for grouped data, the preliminary results of Hall (1985) suggest that, for example, Pearson's X^2 with overlapping cells can be made more powerful than the Cramér-von Mises statistic.

7.3. Comparisons of Discrete and Continuous Test Statistics

In Sections 7.1 and 7.2 the power-divergence statistic (7.2) was compared with test statistics designed specifically for testing goodness-of-fit for continuous data. From these comparisons we draw the following overall conclusions:

(a) The sample quantile statistic X_Q^2 in (7.6) offers no asymptotic or practical advantage over Pearson's X^2 statistic (i.e., (7.2) with $\lambda = 1$). X_Q^2 and X^2 are asymptotically equivalent.
(b) The test statistics based on (first-order) spacings produce comparable distributional results to those under the sparseness assumptions of Section 4.3 (where the number of cells increases with the sample size n). It is possible that a judicious choice of cell boundaries might yield a power-divergence statistic (7.2) with similar asymptotic properties to statistic (7.8).
(c) The results for test statistics based on mth-order spacings indicate that increased power can be obtained by using statistics with overlapping cell boundaries. Further work is needed in this area before recommendations on appropriate cell boundaries can be made for the power-divergence statistic.
(d) Test statistics based on the continuous version of the power-divergence statistic (7.13) are a natural extension of (7.2) (with overlapping cell boundaries) for large k. However the density estimator $f_n(y)$ in (7.14) is the naive histogram and would be better replaced by a more sophisticated kernel estimator. Consequently there may be some practical advantages in using (7.13) with a better kernel estimator, but no comparisons exist between these statistics.
(e) Statistic (7.13) illustrates the major difference between the EDF test statistics and the power-divergence test statistic (7.2). EDF statistics compare distribution functions, whereas the power-divergence statistic accumulates relative *changes* of distribution functions over their supports.

7.4. Diversity and Divergence Measures from Information Theory

The previous sections of this chapter have concentrated on comparisons of the power-divergence family with various other goodness-of-fit statistics defined for continuous data. The discussion aimed at providing a broader perspective on available approaches to testing goodness of fit. This final section provides an alternative perspective by shifting the focus to a different, but closely related set of measures that are used in the field of information theory. These include measures of the diversity (or equivalently the entropy) contained in a single specified distribution and measures of the directed divergence between two (or more) specified distributions. By linking this literature on information measures to a family of directed divergences (directly related to the power-divergence statistic), we develop some descriptive properties and characterizations for these directed divergences. Finally we conclude the section with a discussion of diversity decomposition, which leads to yet another definition of directed divergence based on the diversity between two distributions.

Diversity Indices

Suppose each member of a population is categorized into one of k mutually exclusive cells; a diversity index measures the dispersion of the population among the various cells. In some sense, a diversity index can be considered to be a qualitative analogue to a measure of dispersion for quantitative variables such as the variance or range. Most frequently, the k cells are assumed to represent k different species within the population, and the diversity index represents the diversity of these species. More specifically we define π_1, π_2, $\ldots$, π_k to represent the relative abundance of each species ($\sum_{i=1}^{k} \pi_i = 1$). The diversity index should be a maximum when all species are equally abundant (i.e., $\pi_1 = \pi_2 = \cdots = \pi_k = 1/k$), and should be a minimum (zero) when only one species is present (i.e., $\pi_i = 1$ and $\pi_j = 0$ for all $j \neq i$; i fixed).

The best-known diversity indices include the Shannon index

$$H^0(\boldsymbol{\pi}) = -\sum_{i=1}^{k} \pi_i \log_2(\pi_i)$$

and the Simpson or Gini index

$$H^1(\boldsymbol{\pi}) = 1 - \sum_{i=1}^{k} \pi_i^2.$$

Both indices are special cases of the diversity index of degree-α described by Patil and Taillie (1982) (provided the Shannon index is modified slightly so that natural logarithms are used instead of logarithms to the base 2; it is easy to see that this modification simply multiplies the Shannon index by the positive constant log(2)),

$$H^\alpha(\boldsymbol{\pi}) = \begin{cases} \left(1 - \sum_{i=1}^{k} \pi_i^{\alpha+1}\right) \Big/ \alpha; & \alpha > 0, \\ \lim_{\alpha \to 0} \left(1 - \sum_{i=1}^{k} \pi_i^{\alpha+1}\right) \Big/ \alpha; & \alpha = 0. \end{cases} \tag{7.15}$$

Havrda and Charvát (1967) give an analogous definition for (nonadditive) entropy of order-α.

Other well-known diversity indices include Rényi's (additive) entropy of order-α (Rényi, 1961)

$$R^\alpha(\boldsymbol{\pi}) = \log_2\left(\sum_{i=1}^{k} \pi_i^\alpha\right) \Big/ (1 - \alpha);$$

and one due to Hill (1973), which we discussed in Section 6.2, i.e.,

$$D^\alpha(\boldsymbol{\pi}) = \left(\sum_{i=1}^{k} \pi_i^\alpha\right)^{1/(1-\alpha)}.$$

The properties of these diversity indices are described by many authors, including Mathai and Rathie (1975, 1976), Kempton (1979), Patil and Taillie (1982), Nayak (1985), and van der Lubbe (1986). Rao (1982a, 1982b) considers a different approach to defining the diversity index of a population, which is based on the average "difference" between any two randomly chosen individuals from the population. Rao describes various measures of difference depending on the nature of the population. For example, consider the vector $\mathbf{Y}_1 = (Y_{11}, \ldots, Y_{1m})$ where Y_{1i} can take on only a finite number of values, each with a specified probability. Rao suggests that here Y_{1i} may stand for the type of gene allele at a given locus i on a chromosome. In this case, Rao defines the difference between any two vectors $\mathbf{Y}_1$ and $\mathbf{Y}_2$ to be $d(\mathbf{Y}_1, \mathbf{Y}_2) = m - \sum_{i=1}^{m} \delta_i$, where $\delta_i = 1$ if the ith component of $\mathbf{Y}_1$ and $\mathbf{Y}_2$ agree, and 0 otherwise. The diversity index is then the average of $d(\mathbf{Y}_1, \mathbf{Y}_2)$ over all members $\mathbf{Y}_1, \mathbf{Y}_2$ of the population. When $m = 1$, this index reduces to the Simpson or Gini index $H^1(\boldsymbol{\pi})$. We shall not go into the details of these results here.

Directed Divergence Measures

A diversity index can be considered to measure the divergence between the population distribution $\boldsymbol{\pi} = (\pi_1, \pi_2, \ldots, \pi_k)$ and the uniform distribution $(1/k, 1/k, \ldots, 1/k)$, where an index closer to 0 represents a wider divergence from the uniform. A natural generalization, when considered in this way, is to define a measure of the divergence between two general distributions. One of the first to be considered was Kullback's directed divergence (Kullback, 1959)

$$K(\mathbf{p}:\mathbf{q}) = \sum_{i=1}^{k} p_i \log_2(p_i/q_i), \tag{7.16}$$

where $\mathbf{p}$ and $\mathbf{q}$ are two discrete probability distributions defined on the $(k-1)$-dimensional simplex $\Delta_k = \{\boldsymbol{\pi} = (\pi_1, \pi_2, \ldots, \pi_k): \pi_i \geq 0; i = 1, \ldots, k$ and $\sum_{i=1}^{k} \pi_i = 1\}$. We adopt the convention $p_i \log_2(p_i/q_i) = 0$, when $p_i = 0$ and for any $0 \leq q_i \leq 1$.

The term *directed divergence* indicates that $K(\mathbf{p}:\mathbf{q})$ is not only a measure of the divergence between $\mathbf{p}$ and $\mathbf{q}$, but it also has a directional component since generally $K(\mathbf{p}:\mathbf{q}) \neq K(\mathbf{q}:\mathbf{p})$. Kullback (1959) refers to the symmetrized form

$$J(\mathbf{p}:\mathbf{q}) = K(\mathbf{p}:\mathbf{q}) + K(\mathbf{q}:\mathbf{p}) \tag{7.17}$$

as the divergence and attributes it to Jeffreys (1948).

In Section 3.4 we used $K(\mathbf{p}:\mathbf{q})$ from (7.16) to define the principle of minimum discrimination information, where $K(\mathbf{p}:\mathbf{q})$ was called discrimination information and natural logarithms were used rather than base-2 logarithms (the two differ only by the multiplicative constant log(2)). Berger (1983, p. 134) comments that regretfully $K(\mathbf{p}:\mathbf{q})$ has been referred to by "at least a dozen different names" in the literature.

Two generalizations of $K(\mathbf{p}:\mathbf{q})$ have been discussed extensively in the literature: the additive directed divergence of order-α (Rényi, 1961; also called the information gain of order-α)

$$R^\alpha(\mathbf{p}:\mathbf{q}) = (\alpha - 1)^{-1}\log_2\left[\sum_{i=1}^{k} p_i^\alpha q_i^{1-\alpha}\right]; \qquad \alpha \neq 1, \tag{7.18}$$

and the nonadditive directed divergence of order-α (Rathie and Kannappan, 1972)

$$\tilde{I}^\alpha(\mathbf{p}:\mathbf{q}) = (2^{\alpha-1} - 1)^{-1}\left[\sum_{i=1}^{k} p_i^\alpha q_i^{1-\alpha} - 1\right]; \qquad \alpha \neq 1. \tag{7.19}$$

Both (7.18) and (7.19) collapse to $K(\mathbf{p}:\mathbf{q})$ in (7.16) as $\alpha \to 1$. The basic difference between these two directed divergences is that $R^\alpha(\mathbf{p}:\mathbf{q})$ satisfies the additivity property

$$R^\alpha(\mathbf{p} * \mathbf{r} : \mathbf{q} * \mathbf{s}) = R^\alpha(\mathbf{p}:\mathbf{q}) + R^\alpha(\mathbf{r}:\mathbf{s}), \tag{7.20}$$

where $\mathbf{p}$, $\mathbf{q} \in \Delta_k$; $\mathbf{r}$, $\mathbf{s} \in \Delta_l$; $\mathbf{p} * \mathbf{r}$, $\mathbf{q} * \mathbf{s} \in \Delta_{kl}$ with $\mathbf{p} * \mathbf{r} = (p_1 r_1, p_1 r_2, \ldots, p_1 r_l, p_2 r_1, \ldots, p_k r_l)$ and similarly for $\mathbf{q} * \mathbf{s}$. On the other hand, $\tilde{I}^\alpha(\mathbf{p}:\mathbf{q})$ is called the nonadditive directed divergence because it satisfies

$$\tilde{I}^\alpha(\mathbf{p} * \mathbf{r} : \mathbf{q} * \mathbf{s}) = \tilde{I}^\alpha(\mathbf{p}:\mathbf{q}) + \tilde{I}^\alpha(\mathbf{r}:\mathbf{s}) + c\tilde{I}^\alpha(\mathbf{p}:\mathbf{q})\tilde{I}^\alpha(\mathbf{r}:\mathbf{s}), \tag{7.21}$$

where $c > 0$ is a constant. Throughout this section we assume that the dimension of the probability vectors (e.g., $\mathbf{p} \in \Delta_k$; $\mathbf{r} \in \Delta_l$; $\mathbf{p} * \mathbf{r} \in \Delta_{kl}$) determine the number of terms included in the summations in (7.18) and (7.19).

The Power Divergence

Clearly the nonadditive directed divergence of order-α is directly related (up to a constant depending on α) to a directed-divergence analogue of the power-divergence statistic. Specifically, we define the power divergence for $\mathbf{p}$, $\mathbf{q} \in \Delta_k$ to be

$$I^\lambda(\mathbf{p}:\mathbf{q}) = \frac{1}{\lambda(\lambda+1)}\sum_{i=1}^{k} p_i\left[\left(\frac{p_i}{q_i}\right)^\lambda - 1\right]; \qquad -\infty < \lambda < \infty, \tag{7.22}$$

where the values at $\lambda = 0$ and $\lambda = -1$ are taken to be the continuous limits as $\lambda \to 0$ and $\lambda \to -1$, respectively. We adopt the convention $p_i[(p_i/q_i)^\lambda - 1]/\lambda = 0$ when $p_i = q_i = 0$. In the case $\lambda \to 0$, (7.22) reduces to (7.16) provided the natural rather than base-2 logarithm is used in (7.16). The power divergence (7.22) and the nonadditive directed divergence (7.19) are related by the equation

$$I^\lambda(\mathbf{p}:\mathbf{q}) = \frac{2^\lambda - 1}{\lambda(\lambda+1)}\tilde{I}^{\lambda+1}(\mathbf{p}:\mathbf{q}); \qquad \lambda \neq -1 \tag{7.23}$$

(provided the appropriate limits are taken as $\lambda \to 0$). Therefore most of the properties and characterizations associated with the nonadditive directed divergence $\tilde{I}^{\lambda+1}(\mathbf{p}:\mathbf{q})$ can be applied directly to the power divergence $I^\lambda(\mathbf{p}:\mathbf{q})$. However we shall see, from the nonnegativity property described later, that $I^\lambda(\mathbf{p}:\mathbf{q})$

provides a more natural definition of directed divergence than does $\tilde{I}^{\lambda+1}(\mathbf{p}:\mathbf{q})$, which is not always nonnegative. Furthermore as $\lambda \to -1$, $\tilde{I}^{\lambda+1}(\mathbf{p}:\mathbf{q}) \to 0$ for all $\mathbf{p}, \mathbf{q} \in \Delta_k$, which is a meaningless measure; conversely we saw in Section 2.4 that as $\lambda \to -1$, $I^{\lambda}(\mathbf{p}:\mathbf{q})$ converges to

$$I^{-1}(\mathbf{p}:\mathbf{q}) = \sum_{i=1}^{k} q_i \log(q_i/p_i),$$

which is a well-defined directed divergence.

Generalizations of Directed Divergence

There have been a variety of generalizations of the directed divergence of order-α. These include the definition of a measure of the simultaneous divergence for three probability distributions (Kannappan and Rathie, 1978). Nath (1972), Mathai and Rathie (1975) and Patni and Jain (1977) consider cases where $\sum_{i=1}^{k} p_i < 1$ or $\sum_{i=1}^{k} q_i < 1$. Ali and Silvey (1966) and Csiszár (1978) consider a general class of convex directed divergences of the form $\int_{-\infty}^{\infty} q(x)f(p(x)/q(x))d\mu(x)$ where $p(x)$ and $q(x)$ are density functions, f is a convex function, and μ is a measure. A special case of this general convex class is the power-divergence family, which we consider provides sufficient generality together with a more tangible and descriptive structure through the single real-valued parameter λ.

Properties of the Power Divergence

The properties and characterizations of both the additive and nonadditive directed divergences of order-α are described in detail by Mathai and Rathie (1975, 1976). We shall summarize those relevant to the power divergence $I^{\lambda}(\mathbf{p}:\mathbf{q})$ defined by (7.22). Again we shall assume throughout this section that the number of terms included in the summation in $I^{\lambda}(\mathbf{p}:\mathbf{q})$ depends on the dimension of the vectors $\mathbf{p}$ and $\mathbf{q}$.

(1) *Nonnegativity.* A natural requirement for a measure of divergence is that it take only positive values and that it increase as $\mathbf{p}$ and $\mathbf{q}$ "diverge." In particular, for the power divergence we have

$$I^{\lambda}(\mathbf{p}:\mathbf{q}) \geq 0,$$

with equality if and only if $p_i = q_i$ for all $i = 1, \ldots, k$. This result follows from the strict convexity of the function $\phi(x) = (x^{\lambda+1} - 1)/\lambda(\lambda + 1)$ and Jensen's inequality. On the other hand, the nonadditive directed divergence $\tilde{I}^{\lambda+1}(\mathbf{p}:\mathbf{q})$ satisfies the nonnegativity property only for $\lambda > -1$ (note that $(2^{\lambda} - 1)/\lambda(\lambda + 1)$ in (7.23) is negative for $\lambda < -1$).

(2) *Symmetry.* The value of the power divergence is not affected by a simultaneous and equivalent reordering of the discrete probability masses in

both of the distributions; i.e.,

$$I^\lambda(\mathbf{p}:\mathbf{q}) = I^\lambda(\mathbf{p}':\mathbf{q}'),$$

where $\mathbf{p}' = (p_{a_1}, p_{a_2}, \ldots, p_{a_k})$; $\mathbf{q}' = (q_{a_1}, q_{a_2}, \ldots, q_{a_k})$; and $(a_1, a_2, \ldots, a_k)$ is an arbitrary permutation of the natural order $(1, 2, \ldots, k)$.

(3) *Continuity.* Small changes in the probability distributions under comparison result in only small changes in the power divergence; in other words, $I^\lambda(\mathbf{p}:\mathbf{q})$ is a continuous function in each of its arguments.

(4) *Recursivity or branching principle or grouping property.* The divergence of the two distributions $\mathbf{p}$, $\mathbf{q} \in \Delta_k$ can only decrease when specific intervals are grouped together; i.e.,

$$I^\lambda(\mathbf{p}:\mathbf{q}) = I^\lambda(\mathbf{p}^+ : \mathbf{q}^+) + (p_1 + p_2)^{\lambda+1}(q_1 + q_2)^{-\lambda} I^\lambda(\mathbf{p}^{12} : \mathbf{q}^{12}),$$

where $\mathbf{p}^+ = (p_1 + p_2, p_3, \ldots, p_k) \in \Delta_{k-1}$, $\mathbf{p}^{12} = (p_1/(p_1 + p_2), p_2/(p_1 + p_2)) \in \Delta_2$, and similarly for $\mathbf{q}^+$, $\mathbf{q}^{12}$.

(5) *Expansibility or zero-indifference.* Addition of a zero probability mass to both $\mathbf{p}$ and $\mathbf{q}$ makes no difference to the divergence coefficient; i.e.,

$$I^\lambda(\mathbf{p}:\mathbf{q}) = I^\lambda(\mathbf{p}^\# : \mathbf{q}^\#),$$

where $\mathbf{p}^\# = (p_1, p_2, \ldots, p_k, 0) \in \Delta_{k+1}$, and similarly for $\mathbf{q}^\#$.

(6) *Nonadditivity or logadditivity.* Let $\mathbf{p}$, $\mathbf{q} \in \Delta_k$; $\mathbf{r}$, $\mathbf{s} \in \Delta_l$ and $\mathbf{p} * \mathbf{r}$, $\mathbf{q} * \mathbf{s} \in \Delta_{kl}$, with $\mathbf{p} * \mathbf{r} = (p_1 r_1, p_1 r_2, \ldots, p_1 r_l, p_2 r_1, \ldots, p_k r_l)$, and similarly for $\mathbf{q} * \mathbf{s}$; then

$$I^\lambda(\mathbf{p} * \mathbf{r} : \mathbf{q} * \mathbf{s}) = I^\lambda(\mathbf{p}:\mathbf{q}) + I^\lambda(\mathbf{r}:\mathbf{s}) + \lambda(\lambda + 1) I^\lambda(\mathbf{p}:\mathbf{q}) I^\lambda(\mathbf{r}:\mathbf{s}),$$

which can be rewritten as

$$\log[1 + \lambda(\lambda + 1) I^\lambda(\mathbf{p} * \mathbf{r} : \mathbf{q} * \mathbf{s})] = \log[1 + \lambda(\lambda + 1) I^\lambda(\mathbf{p}:\mathbf{q})] + \log[1 + \lambda(\lambda + 1) I^\lambda(\mathbf{r}:\mathbf{s})].$$

In other words, a special functional of the power divergence is logadditive.

(7) *Strong nonadditivity.* Let $\mathbf{p}_1, \mathbf{p}_2, \ldots, \mathbf{p}_k$ and $\mathbf{q}_1, \mathbf{q}_2, \ldots, \mathbf{q}_k$ all be probability vectors in Δ_l and $\mathbf{p}$, $\mathbf{q} \in \Delta_k$. Define $\mathbf{p}^* = (p_1 \mathbf{p}_1, p_2 \mathbf{p}_2, \ldots, p_k \mathbf{p}_k) \in \Delta_{kl}$ and similarly for $\mathbf{q}^*$; then

$$I^\lambda(\mathbf{p}^* : \mathbf{q}^*) = I^\lambda(\mathbf{p}:\mathbf{q}) + \sum_{i=1}^{k} p_i^{\lambda+1} q_i^{-\lambda} I^\lambda(\mathbf{p}_i : \mathbf{q}_i).$$

(8) *Representation.* Let $f(u, v) = I^\lambda((u, 1 - u) : (v, 1 - v))$ for $0 \leq u \leq 1$, $0 \leq v \leq 1$, which we call the nonadditive directed-divergence function of order-λ. Then the power divergence can be represented as

$$I^\lambda(\mathbf{p}:\mathbf{q}) = \sum_{i=2}^{k} \tilde{p}_i^{\lambda+1} \tilde{q}_i^{-\lambda} f(p_i/\tilde{p}_i, q_i/\tilde{q}_i),$$

where $\tilde{p}_i = \sum_{j=1}^{i} p_j$, $\tilde{q}_i = \sum_{j=1}^{i} q_j$.

(9) *Distance.* In general, a distance d is defined on a set W if for any two elements x, $y \in W$, a real number $d(x, y)$ is assigned that satisfies the following postulates:

(a) $d(x, y) \geq 0$, with equality if and only if $x = y$;
(b) $d(y, x) = d(x, y)$; and
(c) $d(x, z) \leq d(x, y) + d(y, z)$, $x, y, z \in W$ (the triangle inequality).

Consider the special case $W = \Delta_k$ (the $(k-1)$-dimensional simplex) and $d(\mathbf{p}, \mathbf{q}) = I^\lambda(\mathbf{p} : \mathbf{q})$, given by (7.22). Postulate (a) follows for all λ from the nonnegativity property (1). Postulate (b) is satisfied only by $\lambda = -1/2$. This follows by observing that if λ and γ are such that $I^\lambda(\mathbf{p} : \mathbf{q}) = I^\gamma(\mathbf{q} : \mathbf{p})$ for all $\mathbf{p}, \mathbf{q} \in \Delta_k$, then $\gamma = -\lambda - 1$, hence the only value for which $\lambda = \gamma$ is $\lambda = -1/2$. Unfortunately, postulate (c) is not satisfied for $\lambda = -1/2$ as can be verified using the three probability distributions $\mathbf{p} = (0.25, 0.75)$, $\mathbf{q} = (0.75, 0.25)$, and $\mathbf{r} = (0.5, 0.5)$. However the square root of $I^{-1/2}(\mathbf{p} : \mathbf{q})$ satisfies all three postulates (Matusita, 1955) and hence is a true distance measure known as the Matusita distance,

$$M = \left[\sum_{i=1}^{k} (\sqrt{p_i} - \sqrt{q_i})^2 \right]^{1/2}$$

(Matusita, 1954, 1971; Mathai and Rathie, 1972). This is the discrete form of the Hellinger distance discussed in Section 7.2.

Characterizations of the Power Divergence

A variety of characterizations have been proposed in the literature for the additive and nonadditive directed divergences of order-α defined by (7.18) and (7.19). Through suitable adjustments to the normalizing equations, it is possible to use these results to characterize the power divergence $I^\lambda(\mathbf{p} : \mathbf{q})$ based on subsets of properties (1) through (9). Such characterizations are useful for identifying the most important properties associated with the power divergence. Examples of characterizations include:

(a) The continuity property and the nonadditivity (or logadditivity) property, together with two normalizing equations (Sharma and Taneja, 1975).
(b) The symmetry property and recursivity (or grouping) property, together with one normalizing equation (Rathie and Kannappan, 1972; Cressie and Read, 1984).
(c) The nonadditivity property and the representation property, together with two normalizing equations (Mathai and Rathie, 1975).
(d) The nonnegativity property, together with a simple structural requirement on the divergence measure (Rathie, 1973).

Diversity Decomposition

We conclude this section with a discussion of diversity decomposition. This leads to an alternative view of the divergence between two (or more) distributions, and links some recent results of Burbea and Rao (1982), Haberman (1982), and Rao (1982a, 1982b, 1986) with our earlier discussion of diversity indices and measures of divergence.

In the simplest case, consider two communities with probability distributions $\mathbf{p}_1 = (p_{11}, p_{12}, \ldots, p_{1k})$ and $\mathbf{p}_2 = (p_{21}, p_{22}, \ldots, p_{2k})$ and define the mixture of these communities to have probability distribution $\overline{\mathbf{p}} = \omega \mathbf{p}_1 + (1 - \omega)\mathbf{p}_2$; $0 \le \omega \le 1$. The overall diversity of the mixture is denoted $H(\overline{\mathbf{p}})$ for some diversity index H. The diversity index of degree-α defined in (7.15) would be a suitable choice. Rao (1982a, 1982b) describes an alternative approach to obtaining measures of diversity based on the average difference between any two randomly selected individuals from the population (see discussion after (7.15); Rao and Nayak, 1985, and Rao, 1986, call these measures quadratic entropy and derive further properties). For either approach, provided the diversity index (as a functional on Δ_k) is concave, it follows that

$$H(\overline{\mathbf{p}}) \ge \omega H(\mathbf{p}_1) + (1 - \omega) H(\mathbf{p}_2). \tag{7.24}$$

In other words the diversity in the mixture of communities should be no less than the average diversity within the individual communities. This property parallels the well-known analysis-of-variance (ANOVA) result for random variables: that the variance of a mixture is greater than or equal to the weighted average variance of the components. The difference is the between-component variance and is nonnegative.

Consequently $H(\overline{\mathbf{p}})$ can be decomposed into two parts

$$H(\overline{\mathbf{p}}) = [\omega H(\mathbf{p}_1) + (1 - \omega) H(\mathbf{p}_2)] + J(\mathbf{p}_1 : \mathbf{p}_2)$$

where $\omega H(\mathbf{p}_1) + (1 - \omega) H(\mathbf{p}_2)$ measures the within-community diversity and $J(\mathbf{p}_1 : \mathbf{p}_2) = H(\overline{\mathbf{p}}) - \omega H(\mathbf{p}_1) - (1 - \omega) H(\mathbf{p}_2)$ measures the between-community diversity. This decomposition is analogous to the usual ANOVA decomposition of sums of squares for a one-way classification. Hence $J(\mathbf{p}_1 : \mathbf{p}_2)$ can be considered to measure the divergence of $\mathbf{p}_1$ and $\mathbf{p}_2$. In particular, (7.15), the diversity index of degree-α, satisfies the concavity property (7.24). Therefore we can define (using $\alpha = \beta - 1$ in (7.15))

$$J^{\beta}(\mathbf{p}_1 : \mathbf{p}_2) = \frac{1}{(\beta - 1)} \left[\sum_{i=1}^{k} [\omega p_{1i}^{\beta} + (1 - \omega) p_{2i}^{\beta}] - [\omega p_{1i} + (1 - \omega) p_{2i}]^{\beta} \right],$$

which is called J-divergence (when $\omega = 1/2$) by Burbea and Rao (1982), who show it to be convex for $\beta \in [1, 2]$ and discuss a natural generalization to more than two distributions (described further by Rao, 1986). On the other hand Rényi's (additive) entropy of order-α does not satisfy the concavity property (7.24) and so cannot be decomposed in this way. Rao (1982b, 1986) develops this parallel with ANOVA, and defines a more general decomposition for an m-way classification, which he calls the analysis of diversity (ANODIV). These extensions are beyond the scope of this chapter and the interested reader is referred to the papers referenced earlier, as well as Lau (1985) (who uses ANODIV to characterize Rao's approach to diversity), Rao and Nayak (1985) (who derive properties of Rao's quadratic entropy and related measures of cross entropy or directed divergence), and Nayak (1986) (who discusses the sampling distributions of the ANODIV components).

CHAPTER 8

Future Directions

The development of the power-divergence test statistic in the previous chapters suggests interesting and important avenues for further research. In this chapter we outline some specific topics and suggest possible future directions this research might take.

8.1. Hypothesis Testing and Parameter Estimation under Sparseness Assumptions

Under the sparseness assumptions in Section 4.3 we provided efficiency comparisons between the members of the power-divergence family of statistics assuming the equiprobable hypothesis. It was shown that Pearson's X^2 statistic ($\lambda = 1$) is asymptotically optimal for testing against certain local alternatives. The distributional result (4.19) provides a basis for similar efficiency comparisons for the more general null hypothesis (4.18). However, as we pointed out in Section 4.3, the results of Ivchenko and Medvedev (1978) show that uniform optimality will not exist in general. A further problem is the difficulty in interpreting the sequence of null hypotheses defined by (4.18), since the number of parameters may increase without bound as $k \to \infty$. In concluding their comparison of Pearson's X^2 statistic ($\lambda = 1$) and the loglikelihood ratio statistic G^2 ($\lambda = 0$), Ivchenko and Medvedev (1978, p. 774) state "if the basic hypothesis H_0 fixes the probabilities of outcomes of the scheme 'far' from the equiprobable case, then generally speaking it is impossible to assign preference to either [statistic]...." This conclusion is supported by the Monte Carlo studies of Koehler and Larntz (1980). For the more general problem of trying to choose the optimal statistic from the power-divergence family, there are currently no firm guidelines. Consequently it should be a subject of further research to determine the right classes of null models (with unequal probabilities) that need separate consideration.

Models Requiring Parameter Estimation

Under the conditions of Section 4.3, equation (4.19) states that

$$Pr[2n_k I^\lambda(\mathbf{X}_k/n_k : \boldsymbol{\pi}_{0k}) - \mu_k^{(\lambda)})/\sigma_k^{(\lambda)} \geq c] \rightarrow Pr[N(0,1) \geq c], \qquad \text{as } k \rightarrow \infty, \tag{8.1}$$

for $\lambda = 0, 1, 2, 3, \ldots$ and any $c \geq 0$; where $\mu_k^{(\lambda)}$ and $\sigma_k^{(\lambda)}$ depend on the hypothesized probabilities $\boldsymbol{\pi}_{0k}$. When these $\boldsymbol{\pi}_{0k}$ are functions of unknown parameters (which are to be estimated), how can the limit result be used? Koehler (1986) discusses this question with regard to testing loglinear models for contingency tables (Chapter 3), and gives sufficient conditions for asymptotic normality in the case $\lambda = 0$, and when the unknown parameters are replaced by their maximum likelihood estimates. However Koehler notes that bias of the estimated moments is a potential problem for very sparse tables, and that the speed of convergence to the asymptotic distribution appears slow. For testing the goodness of fit of a continuous parametric distribution function with unknown location parameter, Gan (1985) shows that X^2 ($\lambda = 1$) is asymptotically normal when the (location) parameter is estimated via the ungrouped sample median. Fienberg (1979) provides a brief synopsis of an earlier version of Koehler's results (Koehler, 1977), and more recently Dale (1986) has extended these asymptotics for data obtained from product-multinomial sampling. Similar results are expected to hold for general λ, but have not yet been proved.

The Conditional Approach

McCullagh (1985a, 1985b, 1986) proposes an alternative solution to the problem of unknown parameters for large sparse tables. He invokes a conditionality principle, arguing that it is appropriate to condition on a sufficient statistic for the vector of nuisance parameters, thus removing the distributional dependence on these parameters.

The use of conditioning to eliminate nuisance parameters has been questioned; cf. Cochran (1952, p. 326) and Berkson (1978). Under the classical (fixed-cells) assumptions of Section 4.1 there is no conflict, because the limiting distribution of the power-divergence statistic does not depend on the values of the nuisance parameters, but only on the number of linear restrictions placed on the parameters. Consequently the asymptotic conditional distribution is the same as the asymptotic unconditional distribution with BAN-estimated nuisance parameters (Cochran, 1952, p. 326). However, under the sparseness assumptions of Section 4.3, the limiting distribution of the power-divergence statistic does depend on the nuisance parameters. Conditional on sufficient statistics for the nuisance parameters, McCullagh (1985b, 1986) derives the limiting distribution of X^2 ($\lambda = 1$) and G^2 ($\lambda = 0$) under sparseness assumptions similar to those of Morris (1975). Unfortunately these results are not strictly applicable to the general loglinear model environment since they require that the number of estimated parameters remains constant as the number of cells becomes large. McCullagh's results have yet to be extended to the power-divergence statistic, although we conjecture that similar results will hold for general λ.

A Jackknife Estimate of the Standard Deviation of the Test Statistic

In a recent article, Simonoff (1986) suggests another unconditional approach for estimating the standard deviation of the asymptotic normal distribution of X^2 ($\lambda = 1$), and the (conjectured) asymptotic normal distribution of the power-divergence statistic with $\lambda = 2/3$ (under the sparseness assumptions for a multinomial distribution with parameters requiring estimation). Instead of replacing the unknown parameters in the standard deviation formula by their maximum likelihood estimates (Koehler, 1986), Simonoff proposes the use of the bootstrap and jackknife to produce a nonparametric estimate of the limiting standard deviation. He concludes from simulations that the jackknife is the estimator of choice (under the sparseness assumptions considered here) in the sense that it has the smallest bias and variability. Furthermore, his simulations indicate that the jackknife has slightly less bias and variability in the case $\lambda = 2/3$ than $\lambda = 1$. Simonoff points out that when $\lambda = 1$, his unconditional analysis and the conditional analysis of McCullagh (1985a, 1985b, 1986) are asymptotically equivalent since X^2 and the sufficient statistic for the nuisance parameters are asymptotically independent (as proved by McCullagh, 1985b). This result is not true for G^2 ($\lambda = 0$) and is unproved when $\lambda = 2/3$. Further research into this unconditional approach and its relationship to the conditional approach is needed for general λ but the jackknife method for $\lambda = 2/3$ looks promising.

Simonoff (1986) further suggests that in sparse tables it may be preferable to base general parameter estimates on smoothed frequencies rather than on the original frequencies. This can result in improved probability estimation (e.g., Simonoff, 1983; Titterington and Bowman, 1985); however the resulting distributions of the goodness-of-fit statistics are complex (Simonoff, 1985). Consequently further research is needed to understand the implications of such smoothing on the test statistics. Examples of (nonparametric) smoothing of the observed frequencies in sparse tables are contained in Simonoff (1985, 1987) and Burman (1987). Further references for smoothing are given in Bishop et al. (1975, chapter 12).

Small-Sample Accuracy of Asymptotic Significance Levels

Another important consideration is the effect of parameter estimation on the small-sample accuracy of the asymptotic significance levels, under both the classical (fixed-cells) assumptions of Section 4.1 and the sparseness assumptions of Section 4.3. In particular, when testing for independence in two-dimensional contingency tables, various authors have shown that Pearson's X^2 test ($\lambda = 1$) based on the asymptotic chi-squared critical value is accurate in finite samples (for references, see Section 1 of the Historical Perspective). Rudas (1986) shows by simulation that using $\lambda = 2/3$ produces a test statistic for which the small-sample accuracy of the chi-squared significance levels is very similar to that of X^2; on the other hand (in the cases he considers) G^2 tends to reject the null hypothesis too frequently. Bedrick (1987) uses the power-divergence statistic to derive approximate confidence intervals for the ratio of two binomial proportions and concludes that in small samples the best error rates are realized for $\lambda \in [0.67, 1.25]$. Similarly Bedrick and Aragon (1987) obtain approximate confidence intervals and joint confidence regions for the transition probabilities of a stationary first-order Markov chain and conclude that λ should be between 0.50 and 0.75.

These results with parameter estimation support our findings in Chapter 5, obtained there for cases that involved no estimation (and with expected cell frequencies no less than 1). We conjecture that further studies with parameter estimation will confirm that the tail of the exact distribution of the power-divergence statistic $2nI^{\lambda}(\mathbf{X}/n : \hat{\boldsymbol{\pi}})$ (with $\lambda = 1/2, 2/3, 1$) is well approximated by the chi-squared distribution (provided the expected cell frequencies are not too small, e.g., greater than 1). Our general premise is that the statistic with $\lambda = 2/3$ provides an excellent compromise between G^2 ($\lambda = 0$) and X^2 ($\lambda = 1$).

Second-Order Correction Terms

In cases where neither the chi-squared nor the normal asymptotic distribution provide a good approximation to the exact distribution tail function of the power-divergence statistic, it would be important to consider a correction to the normal distribution tail function (8.1). Unlike the expansion derived in Section 5.2 under the classical (fixed-cells) assumptions, the asymptotic normality of the power-divergence statistic under the sparseness assumptions allows direct application of an Edgeworth or Cornish-Fisher expansion (along with an appropriate correction term for discreteness of the data).

McCullagh (1985b, 1986) derives the first three conditional cumulants for both Pearson's X^2 ($\lambda = 1$) and the loglikelihood ratio G^2 ($\lambda = 0$) statistics, and then uses the Cornish-Fisher expansion to derive the critical values for hypothesis testing. McCullagh's approach ignores the potential contribution of a discreteness term similar to (5.10), which raises a question about the accuracy of his results. Yarnold (1972) recommends that, in some situations, a discreteness correction term (as given by (5.10)) is needed to obtain accurate results under the classical (fixed-cells) assumptions; under the sparseness assumptions of Section 4.3 no similar studies have been published, and it is not known if the extra term is needed. McCullagh's calculations could be generalized for other values of λ, but we have not pursued this since the moment-corrected chi-squared distribution tail function (5.7) provides a sufficiently good (and simple) approximation for the cases we consider in Section 5.3.

Further exact and simulation studies are needed to assess the effect of sample size and expected cell frequencies on the adequacy of the normal approximation to the exact distribution function of the power-divergence statistic.

Comparing Models with Constant Difference in Degrees of Freedom

An important alternative approach to hypothesis testing in large sparse contingency tables is proposed by Haberman (1977). He examines the distribution of $X^2 = \sum_{i=1}^{k} (m_{1i} - m_{2i})^2/m_{2i}$ and $G^2 = 2\sum_{i=1}^{k} m_{1i}\log(m_{1i}/m_{2i})$ for comparing two loglinear models H_1 and H_2 when H_2 is a special case of H_1 (i.e., hierarchical models); m_{1i}, m_{2i} represent the expected cell frequencies for cell i under hypotheses H_1 and H_2, respectively. Provided the difference in the number of unknown parameters for the two hypothesized models converges to a finite positive constant d as $k \to \infty$, Haberman shows that both X^2 and G^2 converge to chi-squared random variables with d degrees of freedom (under some regularity conditions on the expected cell frequencies).

Agresti and Yang (1987) perform a series of simulations illustrating the usefulness of these asymptotic results and conclude that more attention needs to be placed on this approach to hypothesis testing in contingency tables. We conjecture that Haberman's result will hold for all the members of the power-divergence family. More research is needed before any practical recommendations can be made.

8.2. The Parameter λ as a Transformation

In Sections 6.5 and 6.6 we discussed the idea of using the parameter λ as a transformation to stabilize the variance and reduce the skewness of individual cell frequencies. The results of those sections suggest that $\lambda = 2/3$ provides a promising transformation (in terms of matching asymptotic moments), and that more research is needed to understand the exact distribution of an individual transformed cell frequency. We now turn briefly to four related topics: choosing a power transformation to achieve additivity; viewing the weighted least squares approach to parameter estimation from the perspective of the power-divergence statistic; using transformations for graphical display; and density estimation.

Transformations to a Linear Model

The principle of generalized minimum power-divergence estimation was introduced in Section 3.5. For the internal constraints problem (ICP), assume we have a fixed probability vector $\mathbf{p}$, an observed frequency vector $\mathbf{x}$, and a set of $s+1$ constraints on the vector of expected frequencies $\mathbf{m}$ (e.g., the marginal constraints of a contingency table)

$$\sum_{i=1}^{k} c_{ij} m_i = \sum_{i=1}^{k} c_{ij} x_i; \qquad j = 0, 1, \ldots, s, \tag{8.2}$$

with $c_{i0} = 1$ for $i = 1, \ldots, k$ (i.e., $m_+ = n$). If there exists a vector $\mathbf{m}^{*(\lambda)}$, for a given value of the parameter λ, which satisfies these constraints and can be written in the additive form

$$\frac{1}{\lambda}\left[\left(\frac{m_i^{*(\lambda)}}{np_i}\right)^{\lambda} - 1\right] = \tau_0 + c_{i1}\tau_1 + c_{i2}\tau_2 + \cdots + c_{is}\tau_s, \tag{8.3}$$

then $\mathbf{m}^{*(\lambda)}$ is called the *generalized minimum power-divergence estimate*. This is the vector (satisfying (8.2)) that is closest to $n\mathbf{p}$ in the sense of minimizing the power-divergence statistic $2I^{\lambda}(\mathbf{m} : n\mathbf{p})$.

The general model (8.3) specializes to the loglinear model by setting $\lambda = 0$, and to the linear model by setting $\lambda = 1$ (Section 3.5). This generalization raises the question as to which scale (i.e., value of λ) is the appropriate one in which to define a linear model. Generally the log scale (i.e., $\lambda = 0$) is used, however other scales have been proposed (e.g., the Lancaster-additive model described in Section 3.5). How can we use the data to find the "correct" scale for analysis (i.e., the scale in which an additive model is most appropriate)? One approach would be to choose the value of λ that minimizes $2I^{\lambda}(\mathbf{m}^{*(\lambda)} : n\mathbf{p})$. The implications of choosing λ in this way require investigation.

Transformations Used in Weighted Least Squares Estimation

The question of finding an appropriate linear scale for analysis arises again in the context of a weighted least squares approach to categorical data analysis (due to Grizzle et al., 1969; and introduced in Section 3.4).

The weighted least squares approach assumes the model for the cell probabilities can be expressed as $\mathbf{g}(\boldsymbol{\pi}) = \boldsymbol{\beta}W'$ for some known function $\mathbf{g}$, where W is the design matrix and $\boldsymbol{\beta}$ is a parameter of constants (see 3.26). The parameter vector $\boldsymbol{\beta}$ is then estimated by weighted least squares, i.e., by minimizing $(\mathbf{g}(\mathbf{x}/n) - \boldsymbol{\beta}W')S^{-}(\mathbf{g}(\mathbf{x}/n) - \boldsymbol{\beta}W')'$ from (3.27), where S^{-} is a generalized inverse of the sample covariance matrix of $\mathbf{g}(\mathbf{x}/n)$. In the literature, generally the $\mathbf{g}$ functions have been limited to linear, log, logistic, or exponential (e.g., Forthofer and Lehnen, 1981).

Finding the appropriate scale for the model is a very important part of data analysis. A change in scale can result in substantial simplification (e.g., fewer interactions are present and need to be modeled; see, e.g., Box et al., 1978, for further discussion of the simplifications achievable by transformation). Consequently, we suggest that adding the power functions

$$\mathbf{g}^{\omega}(\boldsymbol{\pi}) = \left[\frac{\pi_1^{\omega} - 1}{\omega}, \frac{\pi_2^{\omega} - 1}{\omega}, \ldots, \frac{\pi_k^{\omega} - 1}{\omega}\right],$$

(and linear combinations $\mathbf{g}^{\omega}(\boldsymbol{\pi})A$ for known matrices A) would provide much greater flexibility in the choice of an appropriate linear scale for modeling the cell probabilities. Along the lines of Box and Cox (1964), we might estimate ω simultaneously with $\boldsymbol{\beta}$ by plotting the minimum value (with respect to $\boldsymbol{\beta}$) of $(\mathbf{g}^{\omega}(\mathbf{x}/n) - \boldsymbol{\beta}W')S^{-}(\mathbf{g}^{\omega}(\mathbf{x}/n) - \boldsymbol{\beta}W')'$ for each ω. Then choose an interpretable value of ω (e.g., an integer or fraction) for which the curve is near its minimum.

In Section 3.4 it was shown that the weighted least squares estimator based on the power transformation model $\mathbf{g}^{\omega}(\boldsymbol{\pi}) = \boldsymbol{\beta}W'$ will be close to the minimum power-divergence estimator $\hat{\boldsymbol{\pi}}^{(\lambda)}$ obtained by minimizing $2nI^{\lambda}(\mathbf{x}/n : \boldsymbol{\pi})$ where $\lambda = -\omega - 1$ (see (3.25) and (3.28)). Therefore another way to choose the appropriate transformation parameter ω would be to plot the statistic $2nI^{\lambda}(\mathbf{x}/n : \hat{\boldsymbol{\pi}}^{(\lambda)})$ against λ and to choose an interpretable value of λ for which the curve is near its minimum; then set $\omega = -\lambda - 1$. This approach to choosing an appropriate transformation parameter needs further investigation.

Comparing Density Estimates Graphically

Suppose we observe realizations of the random variables $Y_1, Y_2, \ldots, Y_n$, which are independent and identically distributed according to an unknown density function $f(\cdot)$, and we wish to ascertain graphically whether f could be a member of the parametric family $\{g(\cdot; \boldsymbol{\theta}); \boldsymbol{\theta} \in \Theta\}$. For example, g may be the Gaussian density function, $g(y; \boldsymbol{\theta}) = (2\pi\theta_2)^{-1/2} \exp\{-(y - \theta_1)^2/2\theta_2\}$; $-\infty < \theta_1 < \infty$, $\theta_2 > 0$. A common approach has been to compute a histogram f_n from k cells $\{(v_i, v_{i+1}]; i = 1, \ldots, k\}$; where $-\infty < v_1 < v_2 < \cdots < v_{k+1} < \infty$ and the interval $[v_1, v_{k+1}]$ covers the full range of the data. In order for f_n to integrate to 1, care must be taken to normalize by the cell width, giving:

$$f_n(y) \equiv \{\# Y_j\text{s} \in (v_i, v_{i+1}]\}/n(v_{i+1} - v_i); \quad v_i < y \leq v_{i+1}, \quad i = 1, \ldots, k.$$

Then graph the histogram $f_n(y)$ and the maximum likelihood estimate of the parametric density function $g(y; \hat{\boldsymbol{\theta}})$ together as functions of y. This leads to a picture like Figure 8.1, where departures from the parametric family are observed as different heights of the density estimates.

Tukey (1972) reasons that the differences between f_n and g are difficult to assess by eye because the variance of a difference is roughly proportional

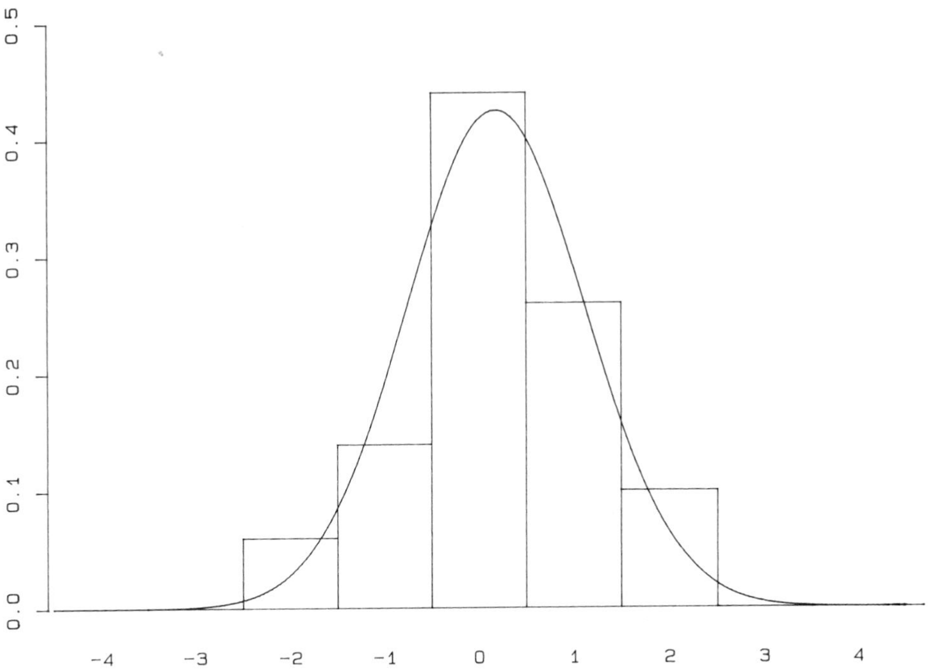

Figure 8.1. Histogram and maximum likelihood estimated density for fifty observations generated from a standard Gaussian distribution. $\hat{\theta}_1 = \bar{y} = 0.1826$, $\hat{\theta}_2 = \sum_{j=1}^{n} (y_j - \bar{y})^2/(n-1) = 0.8808$.

to the height of the density function. Thus in the tails of the density function, small differences may be dismissed as insignificant when in fact they are important. It is easy to see that the square-root transformation yields an approximately constant variance of differences (across cells), which is why Tukey suggests comparing the rootogram $f_n^{1/2}(y)$ to the square-root density $g^{1/2}(y; \hat{\boldsymbol{\theta}})$. One further modification he makes is to move the comparison of histogram to density from a visual inspection of heights to a visual inspection of positive and negative departures from the horizontal axis. He does this by "suspending" the (square-root) histogram box of height $f_n^{1/2}(y)$, where, say $y \in (v_i, v_{i+1}]$, on the estimated (square-root) density function $g^{1/2}((v_i + v_{i+1})/2; \hat{\boldsymbol{\theta}})$ evaluated at the midpoint $(v_i + v_{i+1})/2$ of the cell. Define

$$g_n(y) = \begin{cases} g((v_i + v_{i+1})/2; \hat{\boldsymbol{\theta}}); & \text{for } y \in (v_i, v_{i+1}],\ i = 1, \ldots, k \\ g(y; \hat{\boldsymbol{\theta}}); & \text{for } y \le v_1, \text{ and } y > v_{k+1}. \end{cases} \tag{8.4}$$

Tukey (1972) calls the resultant graphical display a suspended rootogram (further discussion and examples of suspended rootograms are given by Velleman and Hoaglin, 1981). Figure 8.2 shows the suspended rootogram for the data in Figure 8.1 and illustrates the type of information the suspended rootogram displays.

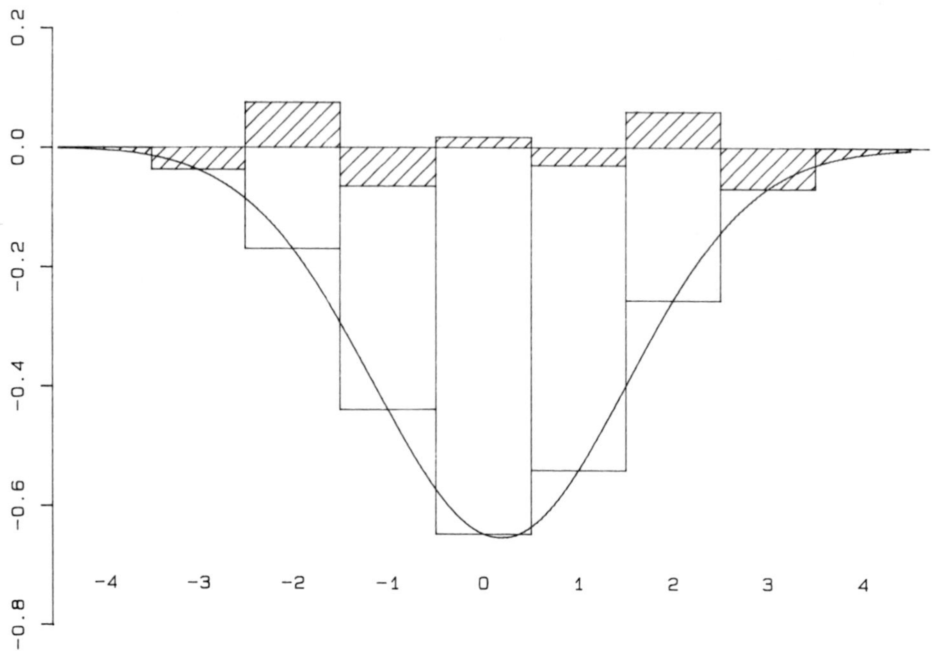

Figure 8.2. Suspended-rootogram plot for the data in Figure 8.1.

The contribution of each cell to the overall estimation of the density function can be seen clearly. Indeed if the values defined by the shaded portion in Figure 8.2 are squared and integrated over the whole line, we obtain

$$\int_{-\infty}^{\infty} [f_n^{1/2}(y) - g_n^{1/2}(y)]^2 dy = -2\int_{-\infty}^{\infty} f_n(y)\left[\left(\frac{f_n(y)}{g_n(y)}\right)^{-1/2} - 1\right]dy$$
$$= (1/2)I^{-1/2}(f_n : g_n);$$

where the general power-divergence formula between the two density functions (i.e., $2nI^\lambda(f_n : g_n)$) is given by equation (7.13) (see also Section 8.3).

The motivation behind modifying the usual histogram comparison plot to a suspended-rootogram plot was to produce a plot that is more sensitive to disagreement of densities over the whole range of comparison. Therefore it is natural to use the ideas suggested by Anscombe (1985), which were developed in Section 6.5. First write

$$2I^\lambda(f_n : g_n) = \frac{2}{\lambda(\lambda+1)}\int_{-\infty}^{\infty} f_n(y)\left[\left(\frac{f_n(y)}{g_n(y)}\right)^{\lambda} - 1\right]dy,$$

as

$$\int_{-\infty}^{\infty} h^\lambda(y)dy,$$

where

$$h^{\lambda}(y) = \frac{2}{\lambda(\lambda+1)}\left\{f_n(y)\left[\left(\frac{f_n(y)}{g_n(y)}\right)^{\lambda} - 1\right] + \lambda[g_n(y) - f_n(y)]\right\}, \quad (8.5)$$

and g_n is given by (8.4). Notice that $h^{\lambda}(y) \geq 0$, and $h^{\lambda}(y) = 0$ when $f_n(y) = g_n(y)$. For $\lambda = 1$,

$$h^1(y) = (f_n(y) - g_n(y))^2/g_n(y);$$

and for $\lambda = -1/2$,

$$h^{-1/2}(y) = 4(f_n^{1/2}(y) - g_n^{1/2}(y))^2,$$

which is proportional to the square of the deviations on the suspended-rootogram plot. Therefore, as a graphical technique to compare two density estimates f_n and g_n, we suggest a plot of

$$h^{\lambda}(y) \text{ versus } y,$$

where $h^{\lambda}(y)$ is given by (8.5). Based on the evidence we have accumulated in the preceding pages of this book, it is our conjecture that $\lambda = 2/3$ in (8.5) will provide a sensitive graphical method of comparison:

$$h^{2/3}(y) = (9/5)[f_n^{5/3}(y)/g_n^{2/3}(y) - (5/3)f_n(y) + (2/3)g_n(y)].$$

Figure 8.3 shows the plot of $h^{2/3}(y)$ versus y (with the parametric density

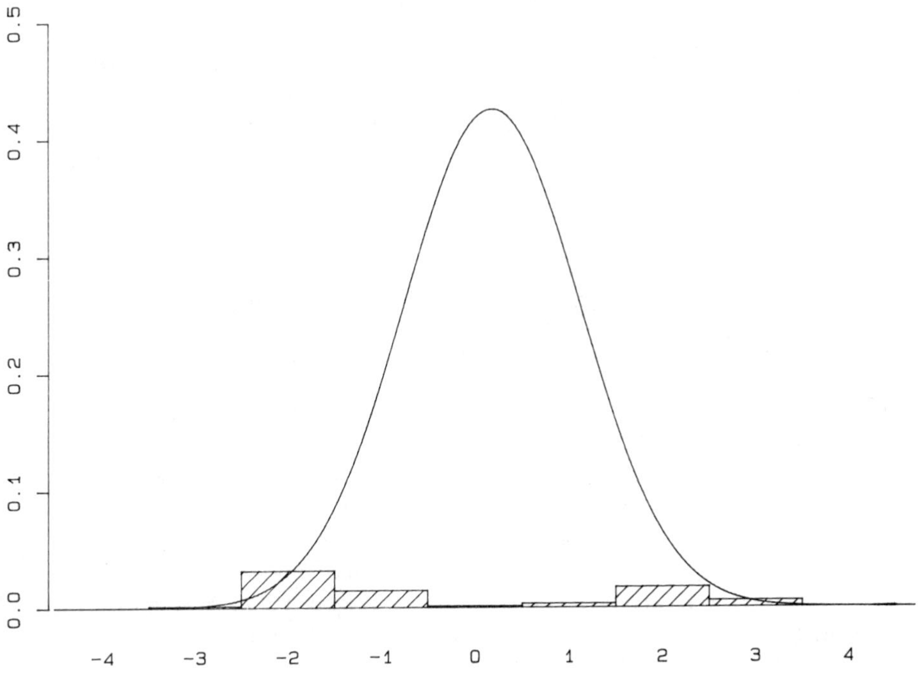

Figure 8.3. Plot of $h^{2/3}(y)$ (with $g(y;\hat{\boldsymbol{\theta}})$ superimposed) for the data in Figure 8.1.

estimate $g(y;\hat{\boldsymbol{\theta}})$ superimposed), based on the same data used in Figures 8.1 and 8.2. Integration over the shaded region yields the measure of difference, $2I^{2/3}(f_n : g_n)$.

In our example data from Figure 8.1, the extreme left and extreme right cells have expectations less than 1 (0.0676 and 0.2347, respectively). Consequently we do not recommend that the chi-squared significance levels (described in Chapters 4 and 5) be used to assess the overall significance of $2I^{2/3}(f_n : g_n)$; this is another area for future research. The results of Section 6.5 (Table 6.7) provide some information about using asymptotics to assess the model fit on a cell-by-cell basis.

Estimating a Density Function

The preceding discussion raises the natural question as to whether the scale $\lambda = 2/3$ would be good also for estimating a density function. In Section 7.2 we introduced the goodness-of-fit measure of Bickel and Rosenblatt (1973), which compares a kernel density estimator f_n to a hypothesized density function f_0 using a continuous analogue of Pearson's X^2. This corresponds to $\lambda = 1$ in the measure

$$2I^{\lambda}(f_n : f_0) = \frac{2}{\lambda(\lambda+1)} \int_{-\infty}^{\infty} f_n(y)\left[\left(\frac{f_n(y)}{f_0(y)}\right)^{\lambda} - 1\right] dy.$$

One possibility then is to estimate the underlying density function f of a random sample, by minimizing $2I^{\lambda}(f_n : f)$ (perhaps including a weight function in the integrand) with respect to f. In the light of our previous discussion, we would recommend choosing $\lambda = 2/3$. If it were desired to impose certain smoothness conditions on the density estimator, a roughness penalty could be added to the criterion in an analogous way to that described by Silverman (1982).

8.3. A Generalization of Akaike's Information Criterion

The Akaike Information Criterion (AIC) was introduced by Akaike (1973) for the purpose of selecting an optimal model from within a set of proposed models (hypotheses). Akaike's criterion is based on minimizing the Kullback directed divergence (Section 7.4) between the true and hypothesized distributions, which we now describe briefly.

Let $\mathbf{y} = (y_1, y_2, \ldots, y_n)$ be a realization of a random vector $\mathbf{Y} = (Y_1, Y_2, \ldots, Y_n)$. Assume the Y_is are independent and identically distributed, each with true density function $g(\cdot;\boldsymbol{\theta}^*)$, where $\boldsymbol{\theta}^* = (\theta_1^*, \theta_2^*, \ldots, \theta_s^*)$ is the true but unknown value of the parameter $\boldsymbol{\theta}$. Now define $\hat{\boldsymbol{\theta}}$ to be the maximum likelihood estimate

(MLE) of $\boldsymbol{\theta}^*$ in some hypothesized set Θ_0, i.e.,

$$l(\hat{\boldsymbol{\theta}}; \mathbf{y}) = \max_{\boldsymbol{\theta} \in \Theta_0} l(\boldsymbol{\theta}; \mathbf{y})$$

where $l(\boldsymbol{\theta}; \mathbf{y}) = \sum_{i=1}^{n} \log(g(y_i; \boldsymbol{\theta}))$ is the loglikelihood function of $\mathbf{y}$. Then the density function $g(\cdot; \hat{\boldsymbol{\theta}})$ is an estimate of $g(\cdot; \boldsymbol{\theta}^*)$. The divergence of $g(\cdot; \hat{\boldsymbol{\theta}})$ from $g(\cdot; \boldsymbol{\theta}^*)$ can be measured by the Kullback directed divergence (described in Section 7.4)

$$K[g(\cdot; \boldsymbol{\theta}^*) : g(\cdot; \hat{\boldsymbol{\theta}})] = \int_{-\infty}^{\infty} g(y; \boldsymbol{\theta}^*) \log\left[\frac{g(y; \boldsymbol{\theta}^*)}{g(y; \hat{\boldsymbol{\theta}})}\right] dy, \tag{8.6}$$

which is a special case (i.e., $\lambda = 0$) of

$$I^{\lambda}[g(\cdot; \boldsymbol{\theta}^*) : g(\cdot; \hat{\boldsymbol{\theta}})] = \frac{1}{\lambda(\lambda + 1)} \int_{-\infty}^{\infty} g(y; \boldsymbol{\theta}^*)\left[\left(\frac{g(y; \boldsymbol{\theta}^*)}{g(y; \hat{\boldsymbol{\theta}})}\right)^{\lambda} - 1\right] dy.$$

Akaike uses the Kullback directed divergence to define the loss function associated with estimating $\boldsymbol{\theta}^*$ by $\hat{\boldsymbol{\theta}}$ (a small value of $K[g(\cdot; \boldsymbol{\theta}^*) : g(\cdot; \hat{\boldsymbol{\theta}})]$ implies $g(\cdot; \hat{\boldsymbol{\theta}})$ is close to $g(\cdot; \boldsymbol{\theta}^*)$). We can rewrite (8.6) as

$$K[g(\cdot; \boldsymbol{\theta}^*) : g(\cdot; \hat{\boldsymbol{\theta}})] = E_Y[\log(g(Y; \boldsymbol{\theta}^*))] - E_Y[\log(g(Y; \hat{\boldsymbol{\theta}}))], \tag{8.7}$$

where Y is a random variable with the same density function $g(\cdot; \boldsymbol{\theta}^*)$ and is independent of the Y_is. Note that the first term on the right-hand side of (8.7) does not depend on the sample. Consequently the second term, which is the expected loglikelihood (evaluated at $\boldsymbol{\theta} = \hat{\boldsymbol{\theta}}$), must decrease as $g(\cdot; \boldsymbol{\theta}^*)$ diverges from $g(\cdot; \hat{\boldsymbol{\theta}})$. However since this quantity depends on the realization $\mathbf{y}$ of the random vector $\mathbf{Y}$, Akaike proposes using minus twice the mean expected loglikelihood,

$$-2 \int E_Y[\log(g(Y; \hat{\boldsymbol{\theta}}))] \exp(l(\boldsymbol{\theta}^*; \mathbf{y})) d\mathbf{y} \tag{8.8}$$

to evaluate the fit of model $g(\cdot; \hat{\boldsymbol{\theta}})$ (a small value of (8.8) implies $g(\cdot; \hat{\boldsymbol{\theta}})$ is a close fit to $g(\cdot; \boldsymbol{\theta}^*)$). Furthermore Akaike shows that if s is the number of (free) parameters estimated by $\hat{\boldsymbol{\theta}}$, then

$$[-2l(\hat{\boldsymbol{\theta}}; \mathbf{y}) + 2s]/n$$

is an unbiased estimate of (8.8) and measures the goodness of fit of the model $g(\cdot; \hat{\boldsymbol{\theta}})$ to $g(\cdot; \boldsymbol{\theta}^*)$. Based on this argument, the AIC is defined to be

$$\text{AIC} = -2l(\hat{\boldsymbol{\theta}}; \mathbf{y}) + 2s, \tag{8.9}$$

and given two hierarchical hypotheses H_0 and H_1 (i.e., H_0 is a special case of H_1), we say H_1 is better than H_0 if $\text{AIC}(H_1) < \text{AIC}(H_0)$. Notice the intuitively appealing presence of the "penalty" $2s$ in the AIC that favors models with fewer fitted parameters. Without it, $-2l(\hat{\boldsymbol{\theta}}; \mathbf{y})$ would always be decreased by choosing H_1 over H_0.

The necessity of avoiding both overparameterization and underparameterization in the model (i.e., the necessity of a parsimony principle) is described in a mathematical context by Azencott and Dacunha-Castelle (1986). Practically, the problem with overparameterizing the model is that while the fit may be excellent for the given data set, the combination of errors from the large number of fitted parameters results in large prediction errors (making it virtually impossible to use the model for predicting the behavior of unobserved data points). On the other hand, too few parameters will result in a model that is incapable of modeling the patterns in the data.

AIC Applied to Categorical Data

Sakamoto and Akaike (1978) and Sakamoto et al. (1986) apply the AIC to the analysis of contingency tables. Following their discussion, let us see how Akaike's criterion compares with the approach based on the power-divergence statistic as described in Section 4.1. Assume $\mathbf{X} = (X_1, X_2, \ldots, X_k)$ has a multinomial distribution $\text{Mult}_k(n, \boldsymbol{\pi})$ where $\boldsymbol{\pi}$ is unknown. Given the null and alternative hypotheses

$$H_0: \boldsymbol{\pi} \in \Pi_0$$
$$H_1: \boldsymbol{\pi} \in \Delta_k,$$

(where $\Delta_k = \{(p_1, p_2, \ldots, p_k): p_i \geq 0; i = 1, \ldots, k \text{ and } \sum_{i=1}^{k} p_i = 1\}$ is the $(k-1)$-dimensional simplex), our traditional hypothesis-testing approach is to reject H_0 if $2nI^{\lambda}(\mathbf{X}/n : \hat{\boldsymbol{\pi}})$ (defined by (2.16)) is "too large" relative to the chi-squared critical value with $k - s - 1$ degrees of freedom. Here $\hat{\boldsymbol{\pi}}$ is the MLE (a BAN estimator) of $\boldsymbol{\pi}$ under H_0.

For the multinomial distribution, the loglikelihood function is (up to a constant)

$$l(\boldsymbol{\pi}; \mathbf{x}) = \sum_{i=1}^{k} x_i \log(\pi_i),$$

consequently the difference between the AICs (8.9) for H_0 and H_1 can be written as

$$\begin{aligned} \text{AIC}(H_0) - \text{AIC}(H_1) &= -2l(\hat{\boldsymbol{\pi}}; \mathbf{x}) + 2s - (-2l(\mathbf{x}/n; \mathbf{x}) + 2(k-1)) \\ &= 2nI^0(\mathbf{x}/n : \hat{\boldsymbol{\pi}}) - 2(k - s - 1). \end{aligned} \tag{8.10}$$

Therefore, checking the sign of $\text{AIC}(H_0) - \text{AIC}(H_1)$ is equivalent to comparing the loglikelihood ratio statistic $2nI^0(\mathbf{x}/n : \hat{\boldsymbol{\pi}})$ with twice its expectation rather than with a fixed percentile of the chi-squared distribution (with $k - s - 1$ degrees of freedom). Table 8.2 gives the probabilities of a chi-squared random variable exceeding twice its expectation. This shows that the significance level at which Akaike's criterion rejects the null hypothesis increases monotonically with s, the number of parameters to be estimated under H_0. (When $k - s - 1 = 7$, the decision based on Akaike's criterion is approxi-

Table 8.2. Probability of a Chi-Squared Random Variable Exceeding Twice Its Expectation

Degrees of freedom									
1	2	3	4	5	6	7	8	9	10
0.16	0.14	0.11	0.09	0.08	0.06	0.05	0.04	0.04	0.03

mately equivalent to that of a traditional test of significance at the 5% level based on $2nI^0(\mathbf{x}/n : \hat{\boldsymbol{\pi}})$.) Sakamoto and Akaike (1978) point out that this property of the AIC increases the tendency to adopt simpler models (which have more degrees of freedom) and hence supports the principle of parsimony (i.e., smaller s) discussed earlier.

BIC—A Consistent Alternative to AIC

Arguing from a Bayesian viewpoint, Schwarz (1978) proposes a related statistic

$$\mathrm{BIC} = 2nI^0(\mathbf{x}/n : \hat{\boldsymbol{\pi}}) - (k - s - 1)\log(n), \tag{8.11}$$

for testing H_0 against H_1; see also Raftery (1986a). As with Akaike's criterion, we accept H_0 if BIC is negative. The BIC criterion has at least two advantages worth considering. First it is a consistent criterion in the contingency table case (Raftery, 1986b). In other words it chooses the correct model with probability 1 as $n \to \infty$; this is not true of the AIC (e.g., Azencott and Dacunha-Castelle, 1986). The second advantage is that the multiplier $\log(n)$ "down-weights" the effective sample size. Traditional significance tests (such as the loglikelihood ratio) can detect any departure from the null model provided the sample size is large enough; this may result in formal rejection of practically useful models such as in the homicide data of Section 6.4.

A natural generalization of AIC and BIC is to replace $\lambda = 0$ in (8.10) and (8.11) with general λ, giving

$$2nI^\lambda(\mathbf{x}/n : \hat{\boldsymbol{\pi}}) - 2(k - s - 1),$$

or

$$2nI^\lambda(\mathbf{x}/n : \hat{\boldsymbol{\pi}}) - (k - s - 1)\log(n).$$

For example, using the homicide data in Section 6.4 with $\lambda = 2/3$, we obtain the values $18.0 - 20.0 = -2.0$ and $18.0 - 97.3 = -79.3$, respectively. Since both these values are negative, we would accept the null model (6.11) using either criterion.

The appeal of using these criteria for model selection is that they automatically penalize models with more parameters. However, viewed from the context of hypothesis-testing theory, the tradeoff of parsimony in favor of power needs investigation. In the case of large sparse tables, Sakamoto and Akaike (1978) note that the AIC needs further refinements.

8.4. The Power-Divergence Statistic as a Measure of Loss and a Criterion for General Parameter Estimation

Suppose that data observed on the random variables $\mathbf{Y} = (Y_1, Y_2, \ldots, Y_n)$ are modeled to have a (joint) distribution function belonging to the parametric family $\{F(\mathbf{y}; \boldsymbol{\theta}); \boldsymbol{\theta} \in \Theta \subset R^k\}$. In order to estimate $\boldsymbol{\theta}$ with some estimator $\hat{\boldsymbol{\theta}} = \boldsymbol{\delta}(\mathbf{Y})$, we need to measure the closeness of $\boldsymbol{\theta}$ to $\boldsymbol{\delta}(\mathbf{Y})$. For $k = 1$, the criterion of squared-error loss,

$$L(\delta, \theta) = (\theta - \delta(\mathbf{Y}))^2,$$

often has been used. For $k > 1$, more caution is needed. The simple-minded generalization

$$\begin{aligned} L(\boldsymbol{\delta}, \boldsymbol{\theta}) &= (\boldsymbol{\theta} - \boldsymbol{\delta}(\mathbf{Y}))(\boldsymbol{\theta} - \boldsymbol{\delta}(\mathbf{Y}))' \\ &= \sum_{i=1}^{k} (\theta_i - \delta_i(\mathbf{Y}))^2 \end{aligned}$$

is meaningless unless all the elements of $\boldsymbol{\theta}$ and $\boldsymbol{\delta}$ have the same units or are unitless. When constructing loss functions, it is important to remember that "apples" should not be added to "oranges," since subsequent comparisons of various estimators through their risks are uninterpretable (e.g., some of the forecasting comparisons in Makridakis et al., 1982).

Instead of squared-error loss, it is sometimes appropriate to consider the squared error as a fraction of the true parameter. Hence, define

$$L^1(\boldsymbol{\delta}, \boldsymbol{\theta}) \equiv \sum_{i=1}^{k} (\theta_i - \delta_i(\mathbf{Y}))^2/\theta_i, \tag{8.12}$$

provided $\theta_1, \theta_2, \ldots, \theta_k$ are all measured in the same units and are all positive.

The loss function (8.12) is one of a number used in the estimation of undercount in the U.S. Decennial Census (e.g., Cressie, 1988). Here θ_i represents the true count in state i, $\delta_i(\mathbf{Y})$ represents the estimated count in that state, and the data $\mathbf{Y}$ are (for example) taken from a postenumeration survey that follows the census typically by several months. It is not only important that the estimated count $\delta_i(\mathbf{Y})$ be close to the true count θ_i, but also that this closeness for a large state does not swamp the contribution to the loss function of smaller states. Equivalently, when (8.12) is expressed in terms of proportions, i.e.,

$$L^1(\boldsymbol{\delta}, \boldsymbol{\theta}) = \sum_{i=1}^{k} \theta_i \left[\frac{\delta_i(\mathbf{Y})}{\theta_i} - 1\right]^2,$$

the contribution to the loss $L^1(\boldsymbol{\delta}, \boldsymbol{\theta})$ when there are two states with the same proportion undercount, should be more for the state with the larger population. This principle is used by the U.S. Congress in its allocation formulas for the number of representatives each state will receive in the House (Spencer, 1985).

Provided the elements of both $\boldsymbol{\theta}$ and $\boldsymbol{\delta}$ are positive, (8.12) can be generalized to the power-divergence loss function

$$L^\lambda(\boldsymbol{\delta}, \boldsymbol{\theta}) \equiv \frac{2}{\lambda(\lambda+1)} \sum_{i=1}^{k} \left\{ \delta_i \left[\left(\frac{\delta_i}{\theta_i} \right)^\lambda - 1 \right] + \lambda[\theta_i - \delta_i] \right\}; \tag{8.13}$$

for $-\infty < \lambda < \infty$, where the cases $\lambda = 0$ and $\lambda = -1$ are defined as the limits of (8.13) as $\lambda \to 0$ and $\lambda \to -1$, respectively. If the estimators are constrained so that $\sum_{i=1}^k \delta_i = \sum_{i=1}^k \theta_i$, or if the δ_is and θ_is are proportions that sum to 1, then

$$L^\lambda(\boldsymbol{\delta}, \boldsymbol{\theta}) = \frac{2}{\lambda(\lambda+1)} \sum_{i=1}^{k} \delta_i \left[\left(\frac{\delta_i}{\theta_i} \right)^\lambda - 1 \right].$$

Fienberg (1986) has suggested this as a family of loss functions to measure the overall differential coverage error in the undercount problem referred to earlier, except that now θ_i and δ_i are, respectively, the true and estimated proportions of state i's population relative to their respective national totals. However, expression (8.13) is the more general one and ensures a positive contribution of each term to the loss function.

Now consider using the loss function (8.13) in the framework of statistical decision theory. In general, suppose the parameter $\boldsymbol{\theta}$ of positive elements has a prior distribution G and it is desired to make a decision about $\boldsymbol{\theta}$ by minimizing the expected loss L, i.e., by minimizing the risk

$$R(G, \boldsymbol{\delta}) = E[L(\boldsymbol{\delta}(\mathbf{Y}), \boldsymbol{\theta})], \tag{8.14}$$

over all possible decisions $\boldsymbol{\delta}$ whose elements are also positive (Ferguson, 1967). It is straightforward to show that the best $\boldsymbol{\delta}$ is obtained by minimizing the expected posterior loss (Ferguson, 1967, p. 44); that is, by minimizing $E[L(\boldsymbol{\delta}(\mathbf{Y}), \boldsymbol{\theta})|\mathbf{Y}]$ over $\boldsymbol{\delta}$. This optimal $\boldsymbol{\delta}$ is called the *Bayes decision function.* Substitution of the power-divergence loss function (8.13) into (8.14) yields a risk we call $R^\lambda(G, \boldsymbol{\delta})$, which is minimized by choosing a $\boldsymbol{\delta}$ that minimizes

$$\frac{2}{\lambda(\lambda+1)} \sum_{i=1}^{k} E\left[\frac{\delta_i^{\lambda+1}}{\theta_i^\lambda} - (\lambda+1)\delta_i + \lambda\theta_i \,|\mathbf{Y} \right].$$

It is straightforward to show that this minimum is achieved by

$$\delta_i(\mathbf{Y}) = [E(\theta_i^{-\lambda}|\mathbf{Y})]^{-1/\lambda}; \qquad i = 1, \ldots, k, \tag{8.15}$$

which is a fractional moment of the posterior distribution, suitably normalized to make the units of the estimator and parameter agree. For example, when $\lambda = 1$, $\delta_i(\mathbf{Y}) = 1/E(\theta_i^{-1}|\mathbf{Y})$; when $\lambda = 0$, $\delta_i(\mathbf{Y}) = \exp[E(\log(\theta_i)|\mathbf{Y})]$; and when $\lambda = -1$, $\delta_i(\mathbf{Y}) = E(\theta_i|\mathbf{Y})$. Thus the conditional expectation of θ_i is an optimal estimator for the particular member $\lambda = -1$, of the power-divergence family of loss functions (8.13). Another way to obtain this as an optimal estimator is to use squared-error loss.

The following illustration gives the optimal estimators for a particular example. Suppose Y is a Poisson random variable with mean θ, which is itself distributed according to a gamma distribution with scale parameter α and shape parameter β. Then the posterior distribution is also a gamma distribution with scale $(\alpha + 1)$ and shape $(\beta + y)$, where y is the realization of Y. Hence

$$\delta(Y) = [E(\theta^{-\lambda}|Y)]^{-1/\lambda} = (\alpha + 1)[\Gamma(\beta + y - \lambda)/\Gamma(\beta + y)]^{-1/\lambda},$$

provided $\beta + y - \lambda > 0$.

Stability of the Bayes Decision Function as λ Varies

Suppose the optimal estimator (8.15) is written as

$$\delta_i(\mathbf{Y}) = g^{-1}[E(g(\theta_i)|\mathbf{Y})],$$

where $g(\theta) = \theta^{-\lambda}$ and g^{-1} is its inverse function. Using the δ-method (Kendall and Stuart, 1969, pp. 231, 244) as a rough way to evaluate leading terms, we see that

$$\delta_i(\mathbf{Y}) \simeq E(\theta_i|\mathbf{Y}),$$

to first order, regardless of λ. To the next order,

$$\begin{aligned}\delta_i(\mathbf{Y}) &\simeq E(\theta_i|\mathbf{Y}) - \frac{\lambda + 1}{2}\frac{\operatorname{var}(\theta_i|\mathbf{Y})}{E(\theta_i|\mathbf{Y})}\\ &= E(\theta_i|\mathbf{Y})\left[1 - \frac{\lambda + 1}{2}\frac{\operatorname{var}(\theta_i|\mathbf{Y})}{E^2(\theta_i|\mathbf{Y})}\right].\end{aligned}$$

Thus, the correction term is linear in $(\lambda + 1)$ and is small when the coefficient of variation of the posterior distribution is small.

Transformed Linear Models

Suppose there is a natural scale $\gamma = -\lambda$ for which the natural parameter θ_i^γ has linear posterior expectation: $E(\theta_i^\gamma|\mathbf{Y}) = \mathbf{a}_i^{(\gamma)}\mathbf{Y}'$. Then the optimal estimator satisfies

$$[\delta_i(\mathbf{Y})]^\gamma = E(\theta_i^\gamma|\mathbf{Y}) = \mathbf{a}_i^{(\gamma)}\mathbf{Y}'.$$

Thus power transforming both parameter and estimator may yield a natural (linear) scale γ. For example $\gamma = 1$ $(\lambda = -1)$ corresponds to the original scale; $\gamma = 0$ $(\lambda = 0)$ corresponds to the log scale; and $\gamma = -1$ $(\lambda = 1)$ corresponds to the reciprocal scale. When $\lambda = -1/2$, $L^{-1/2}(\boldsymbol{\delta}, \boldsymbol{\theta})$ given by (8.13) corresponds to a Hellinger-type loss function. The corresponding $\gamma = -\lambda = 1/2$ yields a square-root scale for the optimal $\delta_i(\mathbf{Y})$. The weighted least squares approach to parameter estimation in contingency tables (due to Grizzle et al., 1969) is an important example of this idea of transforming to a linear scale prior to estimation and analysis (Section 8.2).

General Fitting of Parameters to Data

There is a very simple way that the power-divergence criterion introduced in the preceding discussion can be applied to fitting parametric forms $g(\boldsymbol{\theta})$ to data $\mathbf{Y}$. As a suggestion, instead of fitting parameters $\boldsymbol{\theta}$ by using the ad hoc least-squares criterion, i.e., minimize

$$\sum_{j=1}^{n} (Y_j - g(\boldsymbol{\theta}))^2$$

with respect to $\boldsymbol{\theta}$, try instead the ad hoc family of criteria, i.e., minimize

$$\frac{2}{\lambda(\lambda+1)} \sum_{j=1}^{n} \left\{ Y_j \left[\left(\frac{Y_j}{g(\boldsymbol{\theta})} \right)^{\lambda} - 1 \right] + \lambda[g(\boldsymbol{\theta}) - Y_j] \right\}; \qquad -\infty < \lambda < \infty$$

with respect to $\boldsymbol{\theta}$. This is sensible if the data and the function of parameters are inherently positive. If not, then one possibility is to add a constant to both the Y_js and $g(\boldsymbol{\theta})$ to guarantee positivity.

To illustrate these ideas, suppose $\mathbf{Y} = (Y_1, Y_2, \ldots, Y_n)$ and the Y_js are independent and identically distributed random variables with mean θ; for example, assume $Y_1, Y_2, \ldots, Y_n$ are independent Poisson random variables, with common parameter θ. Consider estimating θ by minimizing the power-divergence statistic

$$\frac{2}{\lambda(\lambda+1)} \sum_{j=1}^{n} \left\{ Y_j \left[\left(\frac{Y_j}{\theta} \right)^{\lambda} - 1 \right] + \lambda[\theta - Y_j] \right\}.$$

Differentiating with respect to θ, and setting the result equal to zero yields the estimator

$$\hat{\theta} = \left[\sum_{j=1}^{n} Y_j^{\lambda+1}/n \right]^{1/(\lambda+1)}.$$

When $\lambda = 0$, $\hat{\theta} = \sum_{j=1}^{n} Y_j/n$, which is the maximum likelihood estimator in the Poisson example. In such a case, where $E(Y_j) = \mathrm{var}(Y_j) = \theta$, it is not unreasonable to estimate θ by weighted least squares, i.e., minimizing

$$\sum_{j=1}^{n} (Y_j - \theta)^2/\theta,$$

which is precisely the power-divergence criterion for $\lambda = 1$. The resulting estimator is $\hat{\theta} = (\sum_{j=1}^{n} Y_j^2/n)^{1/2}$, which is not a function of the sufficient statistic

$\sum_{j=1}^{n} Y_j$. Although $\sum_{j=1}^{n} Y_j$ is the minimum variance unbiased estimator, biased estimators may have smaller mean-squared errors, or as is the case here, smaller relative mean-squared errors.

8.5. Generalizing the Multinomial Distribution

Throughout this book, we have noted that the power-divergence statistic (2.16) with $\lambda = 0$ is equivalent to the loglikelihood ratio statistic G^2 for the multinomial distribution. From this observation, it is natural to ask if there is a more general form of the multinomial distribution (with extra parameters) for which the power-divergence statistic $2nI^{\lambda}(\mathbf{X}/n : \boldsymbol{\pi})$ is the loglikelihood ratio statistic. The answer is not at all immediate.

Historical Perspective: Pearson's X^2 and the Loglikelihood Ratio Statistic G^2

In Chapter 8 we looked towards future directions for research in goodness-of-fit statistics for discrete multivariate data. We now provide some historical perspective on the two "old warriors," X^2 and G^2, which supports our conclusions for the power-divergence family in Chapters 3–7.

Interest, speculation, and some controversy have followed the statistical properties and applications of Pearson's X^2 statistic defined in Section 2.2. The statistic was originally proposed by Karl Pearson (1900) for testing the fit of a model by comparing the set of expected frequencies with the experimental observed frequencies. As a result of the inventiveness of statistical researchers, coupled with advancements in computer technology (allowing more ambitious computations and simulations), an increasingly large cohort of competing tests has become available.

A comprehensive summary of the early development of Pearson's X^2 test is found in Cochran's (1952) review and in Lancaster (1969). Recall

$$X^2 = \sum_{i=1}^{k} \frac{(X_i - n\pi_i)^2}{n\pi_i},$$

where $\mathbf{X} = (X_1, X_2, \ldots, X_k)$ is a random vector of frequencies with $\sum_{i=1}^{k} X_i = n$ and $E(\mathbf{X}) = n\boldsymbol{\pi}$, where $\boldsymbol{\pi} = (\pi_1, \pi_2, \ldots, \pi_k)$ is a vector of probabilities with $\sum_{i=1}^{k} \pi_i = 1$. Pearson (1900) derives the asymptotic distribution of X^2 to be chi-squared with $k - 1$ degrees of freedom when the π_is are known numbers (this result requires the expectations to be large in all the cells; see Section 4.1). In the case where the π_is depend on parameters that need to be estimated, Pearson argued that using the chi-squared distribution with $k - 1$ degrees of freedom would still be adequate for practical decisions. This case was settled finally by Fisher (1924), who gives the first derivation of the correct degrees of freedom, namely, $k - s - 1$ when s parameters are estimated efficiently from the data.

Cochran (1952) provides not only a historical account of the statistical development and applications of Pearson's X^2, but he discusses a variety of competing tests as well. Among these competitors is the loglikelihood ratio test statistic G^2;

$$G^2 = 2 \sum_{i=1}^{k} X_i \log(X_i/n\pi_i),$$

which also has a limiting chi-squared distribution (Wilks, 1938, proves this result in a more general context). When $\mathbf{X}$ is a multinomial random vector with parameters n and $\boldsymbol{\pi}$ (as we shall assume henceforth), G^2 is asymptotically equivalent to Pearson's X^2 statistic (Neyman, 1949; see Section 4.1 for the details of this equivalence). Cochran (1952) notes that to his knowledge there appears to be little to separate these two statistics. As we have seen in the previous chapters, X^2 and G^2 are two special cases in a continuum defined by the power-divergence family of statistics. This family provides an understanding of their similarities and differences and suggests valuable alternatives to them.

Since 1950, the interest in categorical data analysis has renewed discussion on the theory of chi-squared tests in general (i.e., those tests which, under certain conditions, obtain an asymptotic chi-squared distribution), and on how to improve their performance in statistical practice. The general thrust of this research is divided into four areas:

(i) Small-sample comparisons of the test statistics X^2 and G^2 under the null model, when critical regions are obtained from the chi-squared distribution under the classical (fixed-cells) assumptions described in Section 4.1.
(ii) Distributional comparisons of X^2 and G^2 under the sparseness assumptions of Section 4.3.
(iii) Efficiency comparisons made under various assumptions regarding the alternative models and the number of cell boundaries.
(iv) The impact on the test statistics of modifications to the methods of parameter estimation or modifications to the distributional assumptions on the data.

These areas are discussed individually in this Historical Perspective, with emphasis on those results that are relevant to the development of this book.

A variety of general reviews on chi-squared tests have appeared since Cochran's (1952) review. These include Watson (1959), Lancaster (1969), Horn (1977), Fienberg (1979, 1984), and an excellent users review by Moore (1986).

1. Small-Sample Comparisons of X^2 and G^2 under the Classical (Fixed-Cells) Assumptions

When using a test that relies on asymptotic results for computing the critical value, an important question is how well the test (with the asymptotically correct critical value) performs for a finite sample. In this section we use the asymptotic chi-squared distribution for X^2 and G^2 derived under the classical (fixed-cells) assumptions of Section 4.1.

It has long been known that the approximation to the chi-squared distribution for Pearson's X^2 statistic relies on the expected frequencies in each cell of the multinomial being large; Cochran (1952, 1954) provides a complete bibliography of the early discussions regarding this point. A variety of recommendations proliferate in the early literature regarding the minimum expected frequency required for the chi-squared approximation to be reasonably accurate. These values ranged from 1 to 20, and were generally based on individual experience (Good et al., 1970, p. 268, provide an overview of the historical recommendations).

However in the last twenty years, the increase in computer technology has made exact studies and simulations readily available. It has been during this time that the major contributions have been made in understanding the chi-squared approximation in small samples, as we shall now describe.

Suggestions that X^2 approximates a chi-squared random variable more closely than does G^2 (for various multinomial and contingency table models) have been made in enumeration and simulation studies by Margolin and Light (1974), Chapman (1976), Larntz (1978), Cox and Plackett (1980), Koehler and Larntz (1980), Upton (1982), Lawal (1984), Read (1984b), Grove (1986), Hosmane (1986), Koehler (1986), Rudas (1986), Agresti and Yang (1987), Hosmane (1987), and Mickey (1987). The results of Larntz, Upton, and Lawal are of particular interest here because they compare not only X^2 and G^2, but also the Freeman-Tukey statistic T^2;

$$T^2 = \sum_{i=1}^{k} (\sqrt{X_i} + \sqrt{X_i + 1} - \sqrt{4(n\pi_i + 1)})^2.$$

This definition of the Freeman-Tukey statistic differs from F^2 in (2.11) by an order $1/n$ term (Section 6.1; Bishop et al., 1975). Larntz's explanation for the discrepancy in behavior of X^2 to that of T^2 and G^2 is based on the different influence they give to very small observed frequencies. Such observed frequencies increase T^2 and G^2 to a much greater extent than X^2 when the corresponding expected frequencies are greater than 1. In Chapter 6, this phenomenon is discussed in detail for the power-divergence statistic and sheds further light on the differing effect of small expected frequencies on X^2 and G^2 described by Koehler and Larntz (1980).

Other statistics, which are special cases of the power-divergence family, have been considered in the literature. These include the modified loglikelihood ratio statistic or minimum discrimination information statistic for the external constraints problem (Kullback, 1959, 1985; Section 3.5)

$$GM^2 = 2 \sum_{i=1}^{k} n\pi_i \log(n\pi_i/X_i),$$

and the Neyman-modified X^2 statistic (Neyman, 1949)

$$NM^2 = \sum_{i=1}^{k} \frac{(X_i - n\pi_i)^2}{X_i}.$$

While these statistics have been recommended by various authors (e.g., Gokhale and Kullback, 1978; Kullback and Keegel, 1984), there have been no small-sample studies which indicate that they might be serious competitors to X^2 and G^2. The results of Read (1984b) (also Larntz, 1978, Upton, 1982, Lawal, 1984, for T^2, Hosmane, 1987, for F^2) indicate that the exact distributions of these alternative statistics to X^2 and G^2 are less well approximated by the chi-squared distribution than are those of either X^2 or G^2 (Chapters 2 and 5).

The Exact Multinomial, Exact X^2, and Exact G^2 Tests

As an alternative to using the chi-squared distribution for a yardstick to compare chi-squared tests, a number of authors have used the exact multinomial test. The application of this test procedure to draw accuracy comparisons between X^2 and G^2 has provided much confusion and contention in the literature.

The exact multinomial goodness-of-fit test is defined as follows. Given n items that fall into k cells with hypothetical distribution H_0: $\boldsymbol{\pi}_0 = (\pi_{01}, \pi_{02}, \ldots, \pi_{0k})$, where $\sum_{i=1}^{k} \pi_{0i} = 1$, let $\mathbf{x} = (x_1, x_2, \ldots, x_k)$ be the observed numbers of items in each cell. The exact probability of observing this configuration is given by

$$Pr(\mathbf{X} = \mathbf{x}) = n! \prod_{i=1}^{k} \frac{\pi_{0i}^{x_i}}{x_i!}.$$

To test (at an $\alpha 100\%$ significance level) if an observed outcome $\mathbf{x}^*$ came from a population distributed according to H_0, the following four steps are performed:

(1) For every possible outcome $\mathbf{x}$, calculate the exact probability $Pr(\mathbf{X} = \mathbf{x})$.
(2) Rank the probabilities from smallest to largest.
(3) Starting from the smallest rank, add the consecutive probabilities up to and including that associated with $\mathbf{x}^*$; this cumulative probability gives the chance of obtaining an outcome that is no more probable than $\mathbf{x}^*$.
(4) Reject H_0 if this cumulative probability is less than or equal to α.

In a study of the accuracy of the chi-squared approximation for Pearson's X^2 test, Tate and Hyer (1973, p. 836) state, "the chi-square probabilities of X^2 may differ markedly from the exact cumulative multinomial probabilities," and that in small samples the accuracy does not improve as the expected frequencies increase. They note this finding as being "contrary to prevailing opinion." However Radlow and Alf (1975) suggest that these uncharacteristic results are a consequence of the ordering principle that Tate and Hyer use to order the experimental outcomes in the exact multinomial test.

The problem with Tate and Hyer's use of the exact multinomial test is that it orders terms according to their multinomial probabilities instead of according to their discrepancy from H_0 as does X^2. This may not be a good way to define a p-value, and it is justifiable only if events of lower probability are necessarily more discrepant from H_0. According to the experience of Radlow and Alf (1975) and Gibbons and Pratt (1975), this is frequently not true. This point is echoed by Horn (1977, p. 242) when she points out that Tate and Hyer's conclusion that X^2 does not imitate the exact multinomial test for small expected frequencies is "not surprising since the two tests are different tests of the same goodness-of-fit question." The critical values are based on different "distance" measures between the observed and expected frequencies: One is based on probabilities and the other on normalized squared differences.

Radlow and Alf (1975) propose that the appropriate exact test, which should be used to determine the accuracy of the chi-squared approximation for X^2, should be changed. Instead step (2) should read:

(2a) Calculate the Pearson X^2 values under H_0 for each outcome.
(2b) Rank these outcomes from largest to smallest.

The cumulative probability produced at step (3) will now give the chance of obtaining an outcome no "closer" to H_0 than is $\mathbf{x}^*$, where the "distance" measure is the X^2 statistic itself. To obtain an exact test for G^2, the G^2 statistic would replace X^2 in step (2a). Using this revised exact test, which we call the exact X^2 test, Radlow and Alf (1975) obtain results that are much more in accord with the prevailing opinion of independent studies on the accuracy of the chi-squared approximation to X^2 (e.g., Wise, 1963; Slakter, 1966; Roscoe and Byars, 1971, by simulation).

Contrary to the preceding discussion, Kotze and Gokhale (1980) regard the exact multinomial test as optimal for small samples, in the sense that when there is no specific alternative hypothesis it is reasonable to assume that outcomes with smaller probabilities under the null hypothesis should belong to the critical region. The authors proceed by comparing X^2 and G^2 on the basis of "probability ordering" and conclude that G^2 exhibits much closer ranking to the multinomial probabilities. This conclusion was reached previously by Chapman (1976), who states that the exact multinomial probabilities are on average closer to the chi-squared probabilities for G^2 than for X^2. However, Chapman further concludes that the difference between the exact X^2 test probabilities (as revised by steps (2a) and (2b)) and the chi-squared probabilities is usually smaller than the difference between the exact G^2 test probabilities and the chi-squared probabilities.

Our opinion coincides with Radlow and Alf (1975) and the other authors discussed previously. Our computations in Chapter 5 indicate that the chi-squared distribution approximates the distribution of X^2 more closely than the distribution of G^2; furthermore there are members of the power-divergence family "between" X^2 and G^2 for which the chi-squared approximation is even better. We regard X^2 and G^2 as different measures of goodness of fit; it is precisely this difference that makes these tests and the exact multinomial test useful in different situations (e.g., Larntz, 1978; Read, 1984b). This point is discussed with respect to the power-divergence statistic in Chapter 6.

Other small-sample and simulation studies include those of Wise (1963), Lancaster and Brown (1965), Slakter (1966), and Smith et al. (1979), all of whom discuss the case of approximately equal expected frequencies for Pearson's X^2 (Wilcox, 1982, disagrees with some of the calculations of Smith et al.). The consensus of these authors is that here the quality of the chi-squared approximation is more dependent on sample size than on the size of the expected frequencies (see also the discussion of Koehler and Larntz, 1980, later in Section 2). Lewontin and Felsenstein (1965), Haynam and Leone (1965), and Roscoe and Byars (1971) consider the chi-squared approximation for X^2 in contingency table analyses. Lewontin and Felsenstein conclude that X^2 is remarkably robust to expectation size, in the sense that even when expectations less than 1 are present, the approximate chi-squared significance level is quite close to the exact level. Roscoe and Byars (1971) indicate that excellent approximations are achieved using X^2 to test independence for cross-classified categorical data with average expected frequencies as low as 1 or 2. For contingency tables, enumeration of the exact distribution of the test statistic is very computationally intensive and requires fast and efficient algorithms. Verbeek and Kroonenberg (1985) provide a survey of algorithms for $r \times c$ tables with fixed margins.

Further studies relevant to small-sample comparisons of X^2 and G^2 include those of Uppuluri (1969), Good et al. (1970), and Yarnold (1972). Hutchinson (1979) provides a valuable bibliography on empirical studies relating to the validity of Pearson's X^2 test for small expected frequencies; his general conclusion on comparative studies involving X^2, G^2, and T^2 is "the balance of evidence seems to be that Pearson's X^2 is the most robust" (p. 328). Moore (1986, p. 72) also provides recommendations in favor of X^2 for testing against general (unspecified) alternatives.

Finally, Yates (1984) provides some interesting and controversial philosophical disagreements with the comparisons between test statistics reported in much of the recent literature. His conclusions are centered around two main arguments. First, the use of conditional tests is a fundamental principle in the theory of significance tests which has been ignored in the recent literature. Second, the attacks on Fisher's exact test and the continuity corrected X^2 test are due mainly to an uncritical acceptance of the Neyman-Pearson approach to tests of significance and the use of nominal levels instead of the actual probability levels attained. Yates' paper is followed by a controversial discussion of both his examples and conclusions, however, this discussion is beyond the scope of our survey.

Closer Approximations to the Exact Tests

In cases where the chi-squared approximation is considered poor, how might the approximation be improved?

Hoel (1938) provides a second-order term for the large-sample distribution of Pearson's X^2. He concludes from this calculation that the error committed by using only the first-order approximation is much smaller than the neglected terms would suggest. Unfortunately Hoel's theory assumes an underlying continuous distribution; Yarnold (1972) provides the correct second-order term for the discrete multinomial distribution and assesses it, together with four other approximations. However, he still agrees with Hoel's conclusion that the chi-squared approximation can be used with much smaller expectations than previously considered possible.

Jensen (1973) provides bounds on the error of the chi-squared approximation to the exact distribution of X^2. A similar set of bounds are provided for G^2 by Sakhanenko and Mosyagin (1977).

Following Yarnold's (1972) results for X^2, Siotani and Fujikoshi (1984) calculate the appropriate second-order term for the large-sample distributions of G^2 and the Freeman-Tukey statistic F^2 defined in (2.11). This approach has been generalized still further for the power-divergence statistic in Section 5.2 following Read (1984a); however Section 5.3 illustrates that the improvement obtained from these second-order terms can also be obtained from a corrected chi-squared approximation that is far simpler to compute.

Various authors have discussed using moment corrections to obtain approximations closer to the exact distributions of X^2 and G^2. Lewis et al. (1984) provide explicit expressions for the first three moments of X^2 in two-dimensional contingency tables. They indicate that the traditional chi-squared approximation is usable in a wider range of tables than previously suggested; but under certain specified conditions, they recommend using a gamma approximation with the first two moments matched to X^2. In the case of G^2, Lawley (1956) provides an "improved" approximation, which involves multiplying G^2 by a scale factor, giving a statistic with moments equivalent to the chi-squared up to $O(n^{-2})$. This work is extended by Smith et al. (1981) who provide equivalent moments up to $O(n^{-3})$. Cressie and Read's (1984) moment corrections are used in Section 5.1 for the general power-divergence statistic; they provide some worthwhile improvements illustrated in Section 5.3.

Hosmane (1986) considers adding positive constants to all or some cells (i.e., those with zero counts) when testing independence in two-dimensional contingency tables, and then adjusting the resulting statistics X^2 and G^2 to eliminate the dominant bias terms. However, the author's subsequent Monte Carlo study indicates that the distribution of the X^2 statistic has significance levels that are already very close to those of the chi-squared distribution and the adjustment does not improve the accuracy. Indeed, Agresti and Yang (1987) observe that adding constants to the cells before calculating X^2 can play havoc with the distribution of this statistic. On the other hand, the adjustments do provide an improvement for G^2, but the significance levels of the distribution for the adjusted G^2 statistic are still not as close to the chi-squared levels as are those of the unadjusted X^2 statistic. Hosmane's adjustments extend the results of Williams (1976), who provides multipliers to be used with G^2 in contingency table analyses.

A different scaling method for X^2 is used by Lawal and Upton (1984), in which the scaling is chosen to match the upper tail of the exact X^2 distribution with that of the chi-squared distribution (rather than to match the moments). Under certain restrictions, they show this scaling provides accurate results when testing independence in two-dimensional contingency tables with small expectations. Furthermore, they find this approximation preferable to a log-normal approximation, which they had previously recommended (Lawal and Upton, 1980).

In a recent study, Holtzman and Good (1986) assess two corrected chi-squared approximations (and one Poisson approximation) to the exact distribution of Pearson's X^2 statistic for the equiprobable hypothesis. One of the corrections is Cochran's (1942) well-known generalization of Yates' correction. The authors conclude that both corrections generally improve on the uncorrected chi-squared approximation, but the Poisson approximation is less preferable for the cases they consider. This article provides a concise review of the literature on correction factors (including papers not cited here) and an enlightening discussion of how different authors have evaluated such correction factors.

2. Comparing X^2 and G^2 under Sparseness Assumptions

The practical importance of testing hypotheses involving multinomials with many cells, but only a relatively small number of observations, has been emphasized by a number of authors (e.g., Cochran, 1952, p. 330; Fienberg, 1980, pp. 174–175), and was discussed in more detail at the beginning of Section 4.3.

Asymptotic Normality of the Statistics

In his review of Pearson's X^2 test, Cochran (1952, pp. 330–331) points out that when all the expectations are small and X^2 has many degrees of freedom, the distribution of X^2 differs from what would be expected from the chi-squared distribution. Both distributions become approximately normal, however the calculations of Haldane (1937, 1939) show that the variance of X^2 departs noticeably from the variance of the normal approximation to the chi-squared distribution.

Generalizations of this work by Tumanyan (1956), Steck (1957), and Morris (1966) culminated in the landmark paper of Morris (1975), in which he derives the limiting distribution of both X^2 and G^2 under simple null (and certain alternative) hypotheses as both the number of cells $k \to \infty$ and the sample size $n \to \infty$ while n/k remains finite. We refer to these as the sparseness assumptions. Dale (1986) extends these results from single multinomials to include product multinomials. A similar result to Morris', but constrained to the equiprobable null hypothesis, is proved under slightly different regularity conditions by Holst (1972), and corrected by Ivchenko and Medvedev (1978). The details of Morris' and Holst's assumptions and results are provided in Section 4.3, where they are applied to the power-divergence statistic.

While both X^2 and G^2 are asymptotically normal under the sparseness assumptions, it is important to point out that their asymptotic means and variances are no longer equivalent (as they are under the classical fixed-cells assumptions). Koehler and Larntz (1980) suggest that this is predominantly due to the differing influence of very small observed counts on X^2 and G^2. When a given expected frequency is greater than 1, a corresponding observed frequency of 0 or 1 makes a larger contribution to G^2 than to X^2. Therefore, the first two moments of G^2 will be larger than those for X^2 when many expected frequencies are in the range 1 to 5. However, when many expected frequencies are less than 1, the reverse occurs. This idea is expanded on in Chapter 6, where it is used to compare members of the power-divergence family.

Small-Sample Studies

Although discussion of the normal approximation for X^2 has been continuing for nearly forty years, very few empirical studies that assess the adequacy of this approximation in small samples have been published. Most studies have been concerned with the accuracy of the chi-squared approximation, as illustrated by the large number of papers referenced in Section 1 of this Historical Perspective.

The first major Monte Carlo study to examine the relative accuracy of the chi-squared and normal approximations for X^2 and G^2 comes from Koehler and Larntz (1980). For the equiprobable null hypothesis, they conclude that generally the chi-squared approximation for X^2 is adequate with expected frequencies as low as 0.25 when $k \geq 3$, $n \geq 10$, and $n^2/k \geq 10$. Conversely G^2 is generally not well approximated by the chi-squared distribution when $n/k \leq 5$; for $n/k < 0.5$ the chi-squared approximation produces conservative significance levels, and for $n/k > 1$ it produces liberal levels. For $k \geq 3, n \geq 15$, and $n^2/k > 10$, Koehler and Larntz recommend the normal approximation be used for G^2, however their preferred test for the equiprobable hypothesis is X^2 based on the traditional chi-squared approximation. This recommendation is justified further in Section 3, where we shall illustrate the various optimal local power properties of X^2 for this situation.

Under the hypothesis that the cell frequencies come from an underlying Poisson distribution with equal means, Zelterman (1984) uses Berry-Esseen bounds to demonstrate that G^2 and the Freeman-Tukey T^2 are more closely approximated by the normal distribution than is X^2. A small-scale simulation is used to confirm his conclusions. The suggested poor performance of X^2 relative to G^2 in this study does not contradict the results using the chi-squared approximation cited in Section 1, since Zelterman is comparing the performance of both X^2 and G^2 assuming only the normal asymptotic distribution. For equiprobable hypotheses (as assumed by Zelterman) Koehler and Larntz (1980) indicate that the distribution of X^2 is better approximated by the chi-squared than by the normal distribution, and that generally this is a preferable test to using G^2 with the normal approximation. Read (1984b) finds the normal approximation to be much poorer than the chi-squared approximation for both X^2 and G^2 when $10 \leq n \leq 20$ and $2 \leq k \leq 6$ (Section 5.3).

For null hypotheses with unequal cell probabilities, Koehler and Larntz (1980) recommend using G^2 with the normal approximation when most expected frequencies are less than 5, $n \geq 15$, and $n^2/k \geq 10$. The accuracy of the normal approximation for X^2 is seriously affected by a few extremely small expected frequencies, whereas the normal approximation for G^2 is not. With regard to the chi-squared approximation, Koehler and Larntz observe that it produces liberal critical values for X^2 when the null hypothesis has many expected frequencies less than 1; for G^2 it suffers the same shortcomings observed in the equiprobable case.

The Effect of Parameter Estimation

The calculations of Koehler and Larntz (1980) assume that all hypotheses are simple and require no parameter estimation. Koehler (1986) studies the effect of parameter estimation in loglinear models (with closed-form maximum likelihood estimates) on the accuracy of the normal approximation to G^2. He gives sufficient conditions for the asymptotic normality of G^2 with loglinear models; these require not only the number of cells k to increase at a similar rate to the sample size n, but also n must increase faster than the number of parameters estimated. Dale (1986) derives the asymptotic normal distributions of X^2 and G^2 when the underlying sampling model is a product-multinomial with maximum likelihood estimated parameters.

In addition to the asymptotic results for G^2, Koehler (1986) provides a Monte Carlo study of X^2 and G^2 for loglinear models (even though no asymptotic derivations are presented for X^2). The main conclusions are: (a) The chi-squared approximation is generally unacceptable for G^2, but reasonably accurate for X^2 when the expected frequencies are nearly equal; (b) for tables with "both moderately large and very small" expected frequencies, the chi-squared approximation for X^2 can be poor; (c) generally the normal approximation is more accurate for G^2 than for X^2; and (d) substituting the maximum likelihood estimates for the expected frequencies sometimes results in large biases for the moments of G^2. Koehler recommends that less-biased estimators need to be developed.

Conditional Tests

An alternative to the asymptotic approximation of Koehler (1986) is presented by McCullagh (1985a, 1985b, 1986). He argues that it is appropriate to condition on the sufficient statistic for the nuisance parameters; this removes the distributional dependence on the unknown parameters. He then presents a normal approximation for the conditional tests based on X^2 and G^2. These results require the number of estimated parameters to remain fixed as k becomes large. McCullagh (1985a) provides a small-scale simulation study for X^2, which "demonstrate[s] the inadequacy of the normal approximation" (McCullagh, 1986, p. 107). He concludes that an Edgeworth approximation with skewness correction is required for X^2, and expects the same correction will give better results for G^2. However as Koehler (1986) notes, there are still no empirical assessments regarding the accuracy of these Edgeworth approximations relative to the unconditional results of Koehler discussed earlier.

Comparing Models with Constant Difference in Degrees of Freedom

Yet another alternative for testing hypotheses in large sparse contingency tables is to embed the hypotheses of interest in a more general (unsaturated) model, as described by Haberman (1977) (Section 8.1). The (modified) statistics X^2 and G^2, defined to detect the difference between these models, will be approximately chi-squared distributed under certain conditions on the expected frequencies and the hypotheses under comparison. Apart from the analysis of Agresti and Yang (1987), there have been no definitive comparisons of the adequacy of the chi-squared approximation for X^2 and G^2.

The $C(\mathbf{m})$ Distribution

In some situations it may happen that some of the expected frequencies become large with n (according to the classical fixed-cells assumptions) while others remain small (as described under the sparseness assumptions of Section 4.3). Cochran (1942) suggests that in this situation the $C(\mathbf{m})$ distribution should be used, which is defined as follows: Assume that as $n \to \infty$, r expected frequencies remain finite, giving $n\pi_i \to m_i$, $i = 1, \ldots, r$; and $n\pi_i \to \infty$ for $i = r + 1, \ldots, k$. Then the limiting distribution of X^2 is given by the convolution of $\sum_{i=1}^{r} (U_i - m_i)^2/m_i$ and χ^2_{k-r-1} where $\{U_i\}$ are independent Poisson random variables with means $\{m_i\}$, and χ^2_{k-r-1} is a chi-squared random variable with $k - r - 1$ degrees of freedom. The convolution is called the $C(\mathbf{m})$ distribution where $\mathbf{m} = (m_1, m_2, \ldots, m_r)$, and is developed by Yarnold (1970) as a good approximation for X^2 when there are a few small expected frequencies; Lawal

(1980) produces percentage points for the $C(\mathbf{m})$ distribution. Lawal and Upton (1980) recommend using a lognormal approximation for $C(\mathbf{m})$, which in the case of testing independence in contingency tables can be replaced by a scaled chi-squared approximation that is much simpler to use (Lawal and Upton, 1984). The alternative is to use the normal approximation with G^2 as recommended by Koehler and Larntz (1980) and Koehler (1986).

3. Efficiency Comparisons

During the early years, there was much activity devoted to approximating the critical value of X^2 under the null hypothesis. However, Cochran (1952, p. 323) states "the literature does not contain much discussion of the power function of the X^2 test." He attributes this lack of activity to the fact that Pearson's X^2 is used often as an omnibus test against all alternatives, and hence power considerations are not feasible. Some specific alternatives have been considered however, and in this section we present the published results on the efficiency of X^2 and G^2 under both the classical (fixed-cells) assumptions and the sparseness assumptions of Section 4.3.

Classical (Fixed-Cells) Assumptions

Neyman (1949) points out that the X^2 test is consistent; that is, for a nonlocal (fixed) alternative hypothesis, the power of the test tends to 1 as the sample size n increases. This is also true for G^2, implying that the asymptotic power function cannot serve as a criterion for distinguishing between competing tests. To overcome this problem, Cochran (1952) suggests looking at a family of simple local alternatives (i.e., local alternatives requiring no parameter estimation) that converge to the null probability vector as n increases at the rate $n^{-1/2}$. This leads to the comparison criterion called *Pitman asymptotic relative efficiency* (Section 4.2). For such local alternatives, Cochran illustrates (through an argument he attributes to J.W. Tukey) the now well-known result that X^2 attains a limiting noncentral chi-squared distribution (under the classical fixed-cells assumptions; see Section 4.2). The general result for composite local hypotheses (i.e., local alternatives requiring parameter estimation) is proved by Mitra (1958). The identical asymptotic distribution is derived for all members of the power-divergence family in Section 4.2, and indicates that in terms of Pitman asymptotic relative efficiency, no discrimination between X^2, G^2, or any other power-divergence statistic is possible.

In a more general setting, Wald (1943) derives a statistic that (under suitable regularity conditions) possesses asymptotically best average and best constant power over a family of surfaces, and is asymptotically most stringent (Wald, 1943, definitions VIII, X, XII, and equation 162). Subsequently, Cox and

Hinkley (1974, p. 316) show that in the special context of the multinomial distribution, both X^2 and G^2 are asymptotically equivalent to the Wald statistic. From the results of Section 4.2, it follows that the power-divergence statistic inherits these same optimal power properties.

Cohen and Sackrowitz (1975) prove an interesting local optimality property of Pearson's X^2 among a family of statistics that includes the power-divergence statistic. They show that under the classical (fixed-cells) assumptions, X^2 is type-D for testing the equiprobable null hypothesis. This means that among all tests (of the same size) that are locally strictly unbiased, X^2 maximizes the determinant of the matrix of second partial derivatives of the power function evaluated at the null hypothesis. Using the same family of tests as Cohen and Sackrowitz (1975), Bednarski (1980) shows that X^2 is also asymptotically optimally efficient (uniformly minimax) for testing whether the true multinomial probabilities lie within a small neighborhood of the null hypothesis (i.e., testing "ε-validity"). The critical value of the test is obtained from the noncentral chi-squared distribution with noncentrality parameter depending on the size of the neighborhood.

Small-Sample Studies

A small-sample power study of X^2 and G^2 is provided by West and Kempthorne (1971), who plot the exact power curves for both statistics using various composite alternatives. They conclude that there are some regions where X^2 is more powerful than G^2 and vice versa. This conclusion is verified by Koehler and Larntz (1980). On the other hand, Goldstein et al. (1976) perform a series of simulations to compare the power functions of X^2, G^2, and the Freeman-Tukey F^2 (defined in (2.11)) in the case of two- and three-dimensional contingency tables, and they obtain nearly identical results for all three statistics. Wakimoto et al. (1987) calculate some exact powers of X^2, G^2, and F^2 and show that there is a substantial difference between the powers of the test statistics for what we call the bump and dip alternatives (described in Section 5.4). In agreement with our discussion in Section 5.4, they show that X^2 is best for bumps and F^2 is best for dips; G^2 lies between X^2 and F^2.

Two empirical studies by Haber (1980, 1984) indicate that X^2 is more powerful than G^2. In the earlier article he considers inbreeding alternatives to the Hardy-Weinberg null model, and in the later one he considers the general hypothesis of no three-factor interaction in $2 \times 2 \times 2$ contingency tables. It is clear that tests on cross-classified categorical data need special consideration, and since the alternative model plays a big role in any power study, more research in the area of loglinear models is needed before a definitive answer to the question of power can be given.

In a recent Monte Carlo study, Kallenberg et al. (1985) show X^2 and G^2 to have similar power for testing the equiprobable hypothesis using small k. In other cases (i.e., larger k, or hypotheses with unequal cell probabilities), X^2

is better than G^2 for heavy-tailed alternatives and G^2 is better than X^2 for light-tailed alternatives. Throughout their study, the authors adjust G^2 (but not X^2) by a scaling factor to improve the chi-squared significance level approximation in small samples.

Sparseness Assumptions

Under the sparseness assumptions of Section 4.3, the limiting distributions of X^2 and G^2 are different (Section 2 of this Historical Perspective). For the equiprobable null hypothesis, Holst (1972) and Ivchenko and Medvedev (1978) show that X^2 is more powerful than G^2 for testing local alternatives (i.e., X^2 has superior Pitman asymptotic efficiency). Koehler and Larntz (1980) provide Monte Carlo studies that support this conclusion and they remark further, "X^2 is more powerful than G^2 for a large portion of the simplex when k is moderately large" (p. 341). However for null hypotheses with unequal cell probabilities, no general rule applies, and either G^2 or X^2 may be more powerful as illustrated by Ivchenko and Medvedev (1978) and Koehler and Larntz (1980).

Zelterman (1986, 1987) introduces a new statistic

$$D^2 = X^2 - \sum_{i=1}^{k} X_i/(n\pi_i),$$

derived from the loglikelihood ratio statistic for testing a sequence of multinomial null hypotheses against a sequence of local alternatives which are Dirichlet mixtures of multinomials. Zelterman claims that D^2 is not a member of the power-divergence family of goodness-of-fit statistics, however it is easy to show that

$$D^2 = \sum_{i=1}^{k} \frac{(X_i - 1/2 - n\pi_i)^2}{n\pi_i} - k - \sum_{i=1}^{k} \frac{1}{4n\pi_i}.$$

The first term is a "generalized" X^2 statistic, and the last two terms are independent of the data. Thus if the family of power-divergence statistics is expanded to include

$$\sum_{i=1}^{k} h^\lambda(X_i + c, m_i + d),$$

where

$$h^\lambda(u, v) = \frac{2}{\lambda(\lambda + 1)} \left\{ u\left[\left(\frac{u}{v}\right)^\lambda - 1 \right] + \lambda[v - u] \right\},$$

then D^2 is equivalent to the statistic with $\lambda = 1$, $c = -1/2$, $d = 0$. Zelterman shows that D^2 exhibits moderate asymptotic power when the test based on X^2 is biased (i.e., has power smaller than size), and he gives reasons why G^2 and D^2 may have better normal approximations than X^2 when the data are very sparse.

Nonlocal Alternatives

Cochran (1952) points out that using local alternatives is only one way of ensuring that the power of a consistent test is not close to 1 in large samples. Another approach is to use a nonlocal (fixed) alternative but make the significance level $\alpha 100\%$ decrease steadily as n increases. Hoeffding (1965) follows up this approach using the theory of probabilities of large deviations. He shows that by fixing the null and alternative probabilities and letting $\alpha \to 0$ as $n \to \infty$, then G^2 is more powerful than X^2 for simple and some composite alternatives. However, he comments further that there is a need for more study in cases of moderate sample size (other relevant discussions are contained in Oosterhoff and Van Zwet, 1972; Sutrick, 1986).

This approach is similar to one introduced and developed through a series of articles in the 1960s and early 1970s by Bahadur (1960, 1965, 1967, 1971) under the classical (fixed-cells) assumptions. Bahadur (1960, p. 276) states, "... the study (as random variables) of the levels attained when two alternative tests of the same hypothesis are applied to given data affords a method of comparing the performances of the tests in large samples." For a nonlocal (fixed) alternative, Bahadur studies the rate at which the attained significance level of the test statistic tends to 0 as the sample size n becomes large. Bahadur (1971) gives the details for calculating the Bahadur efficiency of X^2 and G^2, which is extended to the power-divergence statistic in Cressie and Read (1984). The most important result is that G^2 attains maximum Bahadur efficiency among all members of the power-divergence family.

These results have been extended by Quine and Robinson (1985), who show that under sparseness assumptions, with n/k finite as $n \to \infty$, G^2 is infinitely superior to X^2 in terms of Bahadur efficiency. This result conflicts with the superior Pitman asymptotic efficiency of X^2 relative to G^2 under the sparseness assumptions of Section 4.3 (see also Kallenberg, 1985). The Bahadur efficiency results need to be viewed in the context of the local optimality properties of X^2 (noted previously for both fixed and increasing k), together with the empirical results, which tend to favor X^2. Studying the rate at which the significance level tends to 0 under a fixed alternative (as the sample size increases) is not in the spirit of traditional hypothesis testing, since there it is the significance level that is held fixed.

Closer Approximations to the Exact Power

The small-sample accuracy of using the noncentral chi-squared distribution to approximate the power function of X^2 is assessed independently by Haynam and Leone (1965), Slakter (1968), and Frosini (1976). All agree that the approximation is not good, and Slakter concludes that the noncentral chi-squared approximation overestimates the power by as much as 20%. Koehler and Larntz (1980) reach a similar conclusion when approximating moderate power levels with the normal distribution (under the sparseness assumptions with no parameter estimation). However, in the case of large sparse contingency tables where parameters need to be estimated, Koehler (1986) comments that there is a clear need for a further assessment of the normal approximation.

Unlike the case for the null hypothesis, very little has been published on closer analytic approximations to the distribution of X^2 or G^2 under alternative hypotheses. Peers (1971) derives a second-order term for the asymptotic distribution of the general loglikelihood ratio statistic for continuous distribution functions. However, these results do not hold when the underlying distribution is multinomial (see also Hayakawa, 1977; Chandra and Ghosh, 1980).

Frosini (1976) proposes a gamma distribution to replace the noncentral chi-squared approximation, and illustrates the improvement through a series of exact studies. Broffitt and Randles (1977) propose a normal distribution to approximate the power function, which they justify for a suitably normalized X^2 statistic when assuming a nonlocal (fixed) rather than local alternative (Section 4.2). Through simulations, they conclude that the normal approximation is more accurate than the noncentral chi-squared distribution when the exact power is large, but the reverse is the case for moderate exact power. This concurs with the previously mentioned comments from Koehler and Larntz (1980) regarding the accuracy of the normal approximation for moderate power levels.

Drost et al. (1987) derive two new large-sample approximations to the power of the power-divergence statistic. These are based on Taylor-series expansions valid for nonlocal alternatives (Broffitt and Randles', 1977, approximation is shown to be a special case after moment correction). Their small-sample calculations show that the traditional noncentral chi-squared approximation is tolerable for X^2 but can be improved using one of their new approximations. For G^2 the noncentral chi-squared approximation can be quite inaccurate, and should be replaced by their new approximations, which perform very well (see also Section 5.4).

Choosing the Cell Boundaries

Depending on the type of data to be analyzed, an experimenter may have to decide on the number of cells k that are to be used in calculating the goodness-of-fit statistic. For example, should some cells be combined? Furthermore, if the data come from an underlying continuous distribution, and are to be partitioned to form a multinomial distribution, the width between the cell boundaries needs to be decided.

Many authors writing on goodness-of-fit tests have proposed that the boundaries should be chosen so the cells are equiprobable, giving $H_0: \pi_{0i} = 1/k; i = 1, \ldots, k$. This choice ensures that X^2 and G^2 (and in general the power-divergence statistic) are unbiased test statistics (Mann and Wald, 1942; Cohen and Sackrowitz, 1975; Sinha, 1976; Spruill, 1977). Subsequent work by Bednarski and Ledwina (1978) and Rayner and Best (1982) indicates that generally (but not always) both X^2 and G^2 (and the power-divergence statistic) are biased test statistics when the cells are not equiprobable. Haberman (1986) illustrates for X^2 that the bias can be quite serious in the case of very sparse contingency tables. In addition, Gumbel (1943) illustrates the potential for strikingly different values of X^2 when using different class intervals with unequal probabilities. All these results help to justify why many authors (including us) concentrate on the equiprobable null hypothesis for small-sample comparisons. However it is important to realize that there are still some disagreements on this issue for certain hypotheses; for example, Kallenberg et al. (1985) argue that using smaller cells in the tails may provide substantial improvement in power for heavy-tailed alternatives (but a loss in power for light-tailed alternatives). For further disagreements on the recommendation to choose equiprobable cells, see Lancaster (1980, p. 118), Ivchenko and Medvedev (1980, p. 545), and Kallenberg (1985).

Assuming the equiprobable cell model, Mann and Wald (1942) derive a formula for choosing the number of cells k (based on sample size n and an $\alpha 100\%$ significance level) so that the power of X^2 is at least $1/2$ for all alternatives no closer than Δ to the equiprobable null probabilities. This result is generalized to other power levels by Harrison (1985). However Williams (1950) shows that for $n > 200$, the value of k given by the Mann-Wald formula can be halved without significant loss of power. The more recent studies of Hamdan (1963), Dahiya and Gurland (1973), Gvanceladze and Chibisov (1979), Best and Rayner (1981), and Quine and Robinson (1985) indicate that the Mann-Wald formula may result in a choice of k that is too large, and reduces the power of the test against specific alternatives.

Oosterhoff (1985, p. 116) illustrates why the viewpoint "that finer partitions are better than coarse ones because less information gets lost" does not hold true for some specific alternatives. Using the example of testing local alternatives with Pearson's X^2 (where X^2 is asymptotically distributed as noncentral chi-squared), Oosterhoff shows that the increase in the number of cells k has two competing asymptotic effects. One effect is an increase in power due to an increase in the noncentrality parameter; the second effect is a decrease in power due to an increase in the variance of X^2. Which of these competing effects is stronger depends on the specific alternative under test.

Best and Rayner (1981) consider testing the null uniform distribution against alternatives that have density functions made up of varying numbers of piecewise linear segments (in particular the uniform density is made up of one piecewise linear segment). They conclude that k need only be just greater than the number of piecewise linear segments specified in their alternative hypothesis. The subsequent calculations of Oosterhoff (1985) and Rayner et al. (1985) indicate that many alternative hypotheses to the equiprobable null model require only small k to achieve reasonable power; however more complicated alternatives involving sharp peaks and dips ("bumps") require larger k. Choosing the "correct" number of cells makes X^2 a much more competitive test statistic than previous results indicate.

Best and Rayner (1985) discuss an alternative approach in which k is set equal to the sample size n, and X^2 is broken down into $k-1$ orthogonal components of which only the first few are used. They illustrate situations in which this approach provides good power for the most frequently encountered alternative hypotheses. Further discussion of the power of orthogonal components of X^2 is provided by Best and Rayner (1987).

4. Modified Assumptions and Their Impact

The effect of modifying the assumptions underlying the goodness-of-fit statistics X^2 and G^2 have been examined from a variety of perspectives. The effects of small expectations and sparseness are two well-known examples, which we discussed in previous sections of this Historical Perspective. We now consider some other modifications.

Estimation from Ungrouped Data

Consider testing the hypothesis that a sample $(y_1, y_2, \ldots, y_n)$ comes from a parametric family $F(y; \boldsymbol{\theta})$ of continuous distributions where $\boldsymbol{\theta} = (\theta_1, \theta_2, \ldots, \theta_s)$ is a vector of nuisance parameters that must be estimated. The traditional goodness-of-fit statistics can be applied to test this hypothesis by grouping the data into k cells with observed frequencies $\mathbf{x} = (x_1, x_2, \ldots, x_k)$, and estimating the expected frequencies of each cell (which depend on $\boldsymbol{\theta}$) with some BAN estimate. The results of Section 4.1 indicate that under H_0, X^2 and G^2 will be asymptotically distributed as chi-squared random variables with $k-s-1$ degrees of freedom. Chernoff and Lehmann (1954) consider the effect of estimating $\boldsymbol{\theta}$ from the original ungrouped observations $(y_1, y_2, \ldots, y_n)$ rather than basing the estimate on the cell frequencies $\mathbf{x}$. They show that the resulting modified X^2 statistic (called the Chernoff-Lehmann statistic) is stochastically larger than the traditional X^2, and has a limiting distribution of the form

$$\sum_{i=1}^{k-s-1} Y_i^2 + \sum_{i=k-s}^{k-1} \lambda_i Y_i^2;$$

where the $\{Y_i\}$ are independent standard normal variates and the $\{\lambda_i\}$ $(0 < \lambda_i < 1)$ may depend on the unknown parameter vector $\boldsymbol{\theta}$. Moore (1986) provides an example of the use of the Chernoff-Lehmann statistic. (In Section 4.1 this modification is generalized to the power-divergence statistic.)

Data-Dependent Cells

A second modification, which has an important impact on the distribution of the Chernoff-Lehmann statistic, is the use of data-dependent cells. Watson

(1959) points out that the choice of cell boundaries for continuous data are typically not fixed but are data-dependent. For example, if the hypothesized distribution were normal, the previous results on the desirability of equiprobable cells may lead us to set up boundaries using the sample mean and variance in order to achieve approximate equiprobable cells.

Roy (1956) and Watson (1957, 1958, 1959) independently observe that with such data-dependent cells, the limit distribution of the Chernoff-Lehmann statistic still takes the form described herein. This extension to random cell boundaries is particularly useful when the family $F(y, \boldsymbol{\theta})$ under test is a location/scale family, since Roy and Watson show the cells can be chosen so that the Chernoff-Lehmann statistic loses its dependence on $\boldsymbol{\theta}$. Dahiya and Gurland (1972, 1973) produce a table of percentage points for the distribution of this statistic, together with some power calculations for testing normality.

When the number of cells increases with n, Gan (1985) derives the asymptotic normality of X^2 when $\boldsymbol{\theta}$ is a one-dimensional location parameter estimated via the ungrouped sample median. This result parallels those under the sparseness assumptions in Section 4.3 (Section 8.1).

Rao and Robson (1974) derive a quadratic form that can be used as an alternative to the Chernoff-Lehmann statistic (with either fixed or random cell boundaries) when the parameter $\boldsymbol{\theta}$ is estimated from the ungrouped data. When testing distributions from the exponential family, the Rao-Robson statistic has the advantage of obtaining a limiting chi-squared distribution with $k - 1$ degrees of freedom regardless of the number of parameters estimated. The review of Moore (1986) provides a useful summary of the Rao-Robson statistic and a further generalization using other estimates of $\boldsymbol{\theta}$.

Watson (1959) points out that random cell boundaries can be used also with the traditional Pearson X^2 statistic, where the parameter vector $\boldsymbol{\theta}$ is estimated from the grouped data. Watson proves that for certain types of data dependence, the test statistic X^2 will still have a limiting chi-squared distribution with degrees of freedom $k - s - 1$, where k is the number of data-dependent cells and s is the dimension of $\boldsymbol{\theta}$. Therefore, after choosing the cell boundaries based on the observed data, the method of choice can be forgotten and Pearson's X^2 statistic can be used to test the hypothesis.

Other relevant articles include that of Moore and Spruill (1975), who provide an account of the distribution theory for a general class of goodness-of-fit statistics comparing both fixed and random cell boundaries. Pollard (1979) generalizes the results on data-dependent methods of grouping for multivariate observations. Moore (1986) provides a succinct summary of these results.

Serially Dependent Data and Cluster Sampling

The effect of serially dependent data on the test statistics is another interesting modification to the traditional assumptions that has received attention recently. When testing the fit of a sequence of observations to a null model for the multinomial probabilities, it is assumed generally that the observations are independent and identically distributed. Gleser and Moore (1985) provide a short survey of the literature regarding serially dependent observations and show that if the successive observations are positively dependent (according to their general definition), then the power-divergence statistic obtains an asymptotic null distribution that is larger than the usual chi-squared distribution. In other words, positive-dependence is confounded with lack of fit. This conclusion is shown to include certain cases of Markov-dependence studied by Altham (1979), Tavaré and Altham (1983), and Tavaré (1983) for testing independence in two-dimensional contingency tables. However Tavaré (1983) illustrates that if one of the two component processes is made up of independent observations and the other is an arbitrary stationary Markov chain, then the X^2 statistic for testing independence is still distributed as a chi-squared random variable with the usual number of degrees of freedom.

Gleser and Moore (1985) point out that another type of dependence is introduced by cluster sampling as described by Cohen (1976), Altham (1976), Brier (1980), Holt et al. (1980), and Rao and Scott (1981). Binder et al. (1984) provide a succinct review of the issues associated with fitting models and testing hypotheses from complex survey designs; they derive the appropriate Wald statistics and find suitable approximations to the null distributions of X^2 and G^2. Koehler and Wilson (1986) extend the results of Brier (1980) for a Dirichlet-multinomial model and study statistics for loglinear models in general survey designs that use some knowledge of this sampling design. A variety of corrections to X^2 and G^2 have been proposed in the papers cited here; other corrections include those of Bedrick (1983), Fay (1985) (using the jackknife), Rao and Scott (1984), and Roberts et al. (1987). Thomas and Rao (1987) study the small-sample properties of some of these corrected statistics under simulated cluster sampling.

For the researcher who suspects serial-dependence but does not wish to model it, the results of Gleser and Moore (1985) are particularly important. They indicate that X^2, G^2, and other members of the power-divergence family may lead to falsely rejecting the null hypothesis.

Overlapping Cell Boundaries

An interesting modification to Pearson's X^2 is discussed by Hall (1985), who considers using overlapping cell boundaries rather than disjoint partitions (Section 7.1). Under certain specified conditions, Hall shows this modified X^2 statistic has superior power to that of the traditional X^2 statistic. A similar modification and conclusion is provided by Ivchenko and Tsukanov (1984) (from a different viewpoint).

Ordered Hypotheses

A modification of the usual hypotheses to incorporate ordered null and alternative hypotheses (e.g., $\pi_1 \leq \pi_2 \leq \cdots \leq \pi_k$) is described by Lee (1987). These hypotheses do not fit into the theory and regularity conditions described in Chapter 4, hence new distributional results must be derived. Lee shows that Pearson's X^2 and the loglikelihood ratio G^2 (and Neyman-modified X^2 statistic) have asymptotically equivalent null distributions consisting of a mixture of chi-squared distributions.

Using Information on the Positions of Observations in Each Cell

Finally we mention the work of Csáki and Vince (1977, 1978), who consider modifying X^2 for grouped continuous data to reflect not only the number of observations in each cell but also their positions relative to the cell boundaries. They derive the limiting distributions of two such modified statistics for both the classical (fixed-cells) assumptions and the sparseness assumptions. Under some specific conditions they show these two statistics to be comparable to, and sometimes more efficient than, X^2; however, the computations for these modified statistics are much more complicated.

Appendix: Proofs of Important Results

A1. Some Results on Rao Second-Order Efficiency and Hodges-Lehmann Deficiency (Section 3.4)

The minimum power-divergence estimator $\hat{\mathbf{m}}^{(\lambda)}$ (from (3.25)) is best asymptotically normal (BAN) for all $\lambda \in (-\infty, \infty)$ (Section A5). Therefore any two estimators $\hat{\mathbf{m}}^{(\lambda_1)}$ and $\hat{\mathbf{m}}^{(\lambda_2)}$ $(\lambda_1 \neq \lambda_2)$ are asymptotically equivalent and are both first-order efficient. However, this asymptotic result gives no information about the rate of convergence.

To compare first-order efficient estimators Rao (1961, 1962) introduces the concept of second-order efficiency, which can be considered a measure of the difference between the Fisher information contained in the sample and the Fisher information contained in the estimator. Subsequently Rao (1963) notes that in the case of multinomial goodness of fit, the second-order efficiency can be derived by calculating the variance of the estimator to order n^{-2} after correcting for the bias to order n^{-1}, as we now show for the power-divergence statistic.

In the following discussion we shall use the parameterization introduced in Section 4.1, where $\boldsymbol{\pi}$ is defined as a function of s parameters $\boldsymbol{\theta} = (\theta_1, \theta_2, \ldots, \theta_s)$ (equation (4.8)), and we shall restrict our attention to the case $s = 1$. It follows from Rao (1963, p. 205) that

$$E[\hat{\theta}^{(\lambda)} - \theta] = b_\lambda(\theta)/n + o(n^{-1})$$

$$= \left[\frac{\lambda}{2\mu_{20}}\left(\sum_{j=1}^{k} \frac{\pi_j'}{\pi_j}\right) - \frac{\lambda\mu_{30} + \mu_{11}}{2\mu_{20}^2}\right] \Big/ n + o(n^{-1}) \qquad \text{(A1.1)}$$

where

$$\mu_{rs} = \sum_{j=1}^{k} \pi_j \left(\frac{\pi_j'}{\pi_j}\right)^r \left(\frac{\pi_j''}{\pi_j}\right)^s,$$

and π_j' and π_j'' are the first and second derivatives of the jth element of $\boldsymbol{\pi}(\theta)$ with respect to θ. In particular μ_{20} is the Fisher information contained in the sample.

Define $\bar{\theta}^{(\lambda)} = \hat{\theta}^{(\lambda)} - b_\lambda(\hat{\theta}^{(\lambda)})/n$, a bias-corrected version of $\hat{\theta}^{(\lambda)}$ resulting from (A1.1). Then, paralleling Rao (1963),

$$\text{var}[\bar{\theta}^{(\lambda)}] = 1/(n\mu_{20}) + (R_1 + \lambda^2 R_2)/n^2 + o(n^{-2}) \tag{A1.2}$$

where

$$R_1 = [-\mu_{20}^3 + \mu_{20}(\mu_{02} - 2\mu_{21} + \mu_{40}) - (\mu_{11}^2/2 + \mu_{30}^2 - 2\mu_{11}\mu_{30})]/\mu_{20}^4,$$

$$R_2 = \left[\frac{\mu_{20}^2}{2} \sum_{j=1}^{k} \left(\frac{\pi_j'}{\pi_j}\right)^2 - \mu_{20}\mu_{40} + \frac{\mu_{30}^2}{2}\right] \Bigg/ \mu_{20}^4,$$

and R_1 and R_2 are nonnegative.

From (A1.2) it is clear that the variance of this corrected minimum power-divergence estimator $\bar{\theta}^{(\lambda)}$ is smallest when $\lambda = 0$ (the MLE); and as $|\lambda|$ increases, the variance of the estimator increases. Furthermore, the (Rao) second-order efficiency of the estimator $\hat{\theta}^{(\lambda)}$ can be calculated as

$$E_2 = Q_1 + \lambda^2 Q_2, \tag{A1.3}$$

where $Q_1 = \mu_{20}^2 R_1 - \mu_{11}^2/2\mu_{20}^2$ and $Q_2 = \mu_{20}^2 R_2$, for R_1 and R_2 defined in (A1.2) (Rao, 1962, table 1). Consequently, according to the criteria of bias-corrected second-order variance (A1.2) and (Rao) second-order efficiency (A1.3), the MLE ($\lambda = 0$) is optimal among the family of minimum power-divergence estimators.

In an interesting and controversial paper, Berkson (1980) calls into question Rao's concept of second-order efficiency. In particular, he questions the appropriateness of calculating the mean-squared error (MSE) of a bias-corrected estimator rather than the MSE of the original estimator. Berkson argues that if $\hat{\theta}$ is the estimator used for θ, then it is the MSE of $\hat{\theta}$ that is of interest and not the MSE of some modified statistic that perhaps is not even calculable for a given data set (Berkson, 1980, p. 462). We agree with Berkson's conclusion (and Parr, 1981; Causey, 1983; Harris and Kanji, 1983), and now consider an alternative method of comparing first-order efficient estimators.

Rao (1963) has set up the general machinery from which we can calculate the MSE of the uncorrected minimum power-divergence estimator $\hat{\theta}^{(\lambda)}$, giving

$$E[\hat{\theta}^{(\lambda)} - \theta]^2 = \text{var}[\bar{\theta}^{(\lambda)}] + \{(2/\mu_{20})b_\lambda'(\theta) + b_\lambda^2(\theta)\}/n^2 + o(n^{-2}) \tag{A1.4}$$

where $b_\lambda'(\theta)$ is the derivative of $b_\lambda(\theta)$. From (A1.1) and (A1.2) it follows that

$$E[\hat{\theta}^{(\lambda)} - \theta]^2 = \frac{1}{n\mu_{20}} + \frac{1}{2n^2}\left\{\frac{-2}{\mu_{20}} + \frac{1}{\mu_{20}^2}\left[2\lambda \sum_{j=1}^{k} \frac{\pi_j''}{\pi_j} + (\lambda^2 - 2\lambda) \sum_{j=1}^{k} \left(\frac{\pi_j'}{\pi_j}\right)^2 \right.\right.$$
$$\left. + \frac{\lambda^2}{2}\left(\sum_{j=1}^{k} \frac{\pi_j'}{\pi_j}\right)^2\right] + \frac{1}{\mu_{20}^3}\left[-2\sum_{j=1}^{k} \frac{\pi_j'\pi_j'''}{\pi_j} - 2(3\lambda + 1)\mu_{21}\right.$$
$$\left. - 2(\lambda^2 - 2\lambda - 1)\mu_{40} - 5\lambda\mu_{11} \sum_{j=1}^{k} \frac{\pi_j'}{\pi_j} - (\lambda^2 - 2\lambda)\mu_{30} \sum_{j=1}^{k} \frac{\pi_j'}{\pi_j}\right]$$
$$\left. + \frac{1}{\mu_{20}^4}\left[\frac{15}{2}\mu_{11}^2 + \left(\frac{3}{2}\lambda^2 - 4\lambda - 2\right)\mu_{30}^2 + 9\lambda\mu_{11}\mu_{30}\right]\right\} + o(n^{-2}). \tag{A1.5}$$

This result can be derived also from Ponnapalli (1976), who calculates the second-order variance of estimators based on minimizing $\sum_{j=1}^{k} x_j\psi(n\pi_j(\theta)/x_j)$ for a suitable function ψ and real-valued θ. Defining $\psi(x) = x^{-\lambda}/\lambda(\lambda + 1)$, (A1.5) follows from Ponnapalli's theorem 1 by noting $E[\hat{\theta}^{(\lambda)} - \theta]^2 = \text{var}[\hat{\theta}^{(\lambda)}] + b_\lambda^2(\theta)/n^2 + o(n^{-2})$.

The leading term in (A1.5) does not depend on λ, so the choice of a good estimator will depend on the term of $O(n^{-2})$. To compare two minimum power-divergence estimators $\hat{\theta}^{(\lambda_1)}$ and $\hat{\theta}^{(\lambda_2)}$, we look at the difference of their second-order asymptotic MSEs (A1.5), which eliminates the equivalent first-order term $1/(n\mu_{20})$. This difference is proposed by Hodges and Lehmann (1970) to compare estimators of equal efficiency to first order, and after division by an appropriate factor, is referred to as the estimators' deficiency. From (A1.5) the Hodges-Lehmann deficiency of $\hat{\theta}^{(\lambda_1)}$ with respect to $\hat{\theta}^{(\lambda_2)}$ becomes

$$D_{HL}[\hat{\theta}^{(\lambda_1)}, \hat{\theta}^{(\lambda_2)}] = \frac{1}{2\mu_{20}}\left[2(\lambda_1 - \lambda_2) \sum_{j=1}^{k} \frac{\pi_j''}{\pi_j} + (\lambda_1^2 - \lambda_2^2 - 2(\lambda_1 - \lambda_2)) \sum_{j=1}^{k} \left(\frac{\pi_j'}{\pi_j}\right)^2\right.$$
$$\left. + \frac{\lambda_1^2 - \lambda_2^2}{2}\left(\sum_{j=1}^{k} \frac{\pi_j'}{\pi_j}\right)^2\right] + \frac{1}{2\mu_{20}^2}\left[6(\lambda_2 - \lambda_1)\mu_{21} + 2(\lambda_2^2 - \lambda_1^2\right.$$
$$- 2(\lambda_2 - \lambda_1))\mu_{40} + 5(\lambda_2 - \lambda_1)\mu_{11} \sum_{j=1}^{k} \frac{\pi_j'}{\pi_j} + (\lambda_2^2 - \lambda_1^2$$
$$\left. - 2(\lambda_2 - \lambda_1))\mu_{30} \sum_{j=1}^{k} \frac{\pi_j'}{\pi_j}\right] + \frac{1}{2\mu_{20}^3}\left[\left(\frac{3}{2}(\lambda_1^2 - \lambda_2^2)\right.\right.$$
$$\left.\left. - 4(\lambda_1 - \lambda_2)\right)\mu_{30}^2 + 9(\lambda_1 - \lambda_2)\mu_{11}\mu_{30}\right]. \tag{A1.6}$$

A justification for using Hodges-Lehmann deficiency follows from the fact that if n and m_n are defined as the sample sizes required for $\hat{\theta}^{(\lambda_1)}$ and $\hat{\theta}^{(\lambda_2)}$, respectively, to have MSEs satisfying the inequality

$$\text{MSE}_{m_n+1}[\hat{\theta}^{(\lambda_2)}] < \text{MSE}_n[\hat{\theta}^{(\lambda_1)}] \leq \text{MSE}_{m_n}[\hat{\theta}^{(\lambda_2)}],$$

then

$$D_{HL}[\hat{\theta}^{(\lambda_1)}, \hat{\theta}^{(\lambda_2)}] = \lim_{n\to\infty} (n - m_n).$$

(Note that $m_n/n \to 1$ since $\hat{\theta}^{(\lambda_1)}$ and $\hat{\theta}^{(\lambda_2)}$ are equally efficient to first order.)

For the special case of the MLE ($\lambda_2 = 0$), (A1.6) simplifies to

$$D_{HL}[\hat{\theta}^{(\lambda)}, \hat{\theta}^{(0)}] = \lambda^2 T_1/2\mu_{20}^3 - \lambda T_2/\mu_{20}^3, \tag{A1.7}$$

where

$$T_1 = \mu_{20}^2\left[\sum_{j=1}^{k}\left(\frac{\pi_j'}{\pi_j}\right)^2 + \frac{1}{2}\left(\sum_{j=1}^{k}\frac{\pi_j'}{\pi_j}\right)^2\right] - \mu_{20}\left[2\mu_{40} + \mu_{30}\sum_{j=1}^{k}\frac{\pi_j'}{\pi_j}\right] + \frac{3}{2}\mu_{30}^2$$

and

$$T_2 = \mu_{20}^2\left[-\sum_{j=1}^{k}\frac{\pi_j''}{\pi_j} + \sum_{j=1}^{k}\left(\frac{\pi_j'}{\pi_j}\right)^2\right] + \mu_{20}\left[3\mu_{21} - 2\mu_{40}\right.$$

$$\left. + \left(\frac{5}{2}\mu_{11} - \mu_{30}\right)\sum_{j=1}^{k}\frac{\pi_j'}{\pi_j}\right] + 2\mu_{30}^2 - \frac{9}{2}\mu_{11}\mu_{30}.$$

Using (A1.7) the following results show that the MLE is no longer uniformly optimal as it is for the criterion of (Rao) second-order efficiency.

Theorem A1.1. *For $k > 2$, $\hat{\theta}^{(\lambda)}$ is least deficient for θ when $\lambda = T_2/T_1$ (defined by* (A1.7)*).*

Proof. Under the regularity conditions of Section A5 it is straightforward to show that $T_1 \geq 0$, with equality if and only if $k = 2$. Therefore when $k > 2$, it is clear from (A1.7) that $D_{HL}[\hat{\theta}^{(\lambda)}, \hat{\theta}^{(0)}]$ is minimized when $\lambda = T_2/T_1$. Finally, from the transitivity of deficiency, it follows that $D_{HL}[\hat{\theta}^{(\lambda_1)}, \hat{\theta}^{(\lambda_2)}] \geq 0$ if and only if $D_{HL}[\hat{\theta}^{(\lambda_1)}, \hat{\theta}^{(0)}] \geq D_{HL}[\hat{\theta}^{(\lambda_2)}, \hat{\theta}^{(0)}]$, which proves the theorem. □

The next result follows immediately.

Corollary A1.1. *When $k > 2$, (a) the MLE ($\lambda = 0$) is least deficient for θ if and only if $T_2 = 0$; and (b) the minimum chi-squared estimator ($\lambda = 1$) is least deficient for θ if and only if $T_1 = T_2$.* □

Theorem A1.1 indicates that the optimal choice of λ (in terms of deficiency) is dependent on the unknown parameter θ. One way to overcome this problem is to choose $\lambda = \lambda^*$ according to the minimax criterion

$$D_{HL}[\hat{\theta}^{(\lambda^*)}, \hat{\theta}^{(0)}] = \min_{\lambda}\max_{\theta} D_{HL}[\hat{\theta}^{(\lambda)}, \hat{\theta}^{(0)}].$$

Alternatively by calculating $\lambda = T_2/T_1$ for each possible value of θ, we can produce regions of "near optimal" λ-values to be used when the true θ is expected to lie in a specific region of the parameter space. For example, consider the four-cell multinomial distribution (i.e., $k = 4$) with hypothetical probabilities

$$\pi_1(\theta) = \pi_2(\theta) = \theta/2, \qquad \pi_3(\theta) = \pi_4(\theta) = (1-\theta)/2.$$

From (3.25) the minimum power-divergence estimator $\hat{\theta}^{(\lambda)}$ can be shown to be

$$\hat{\theta}^{(\lambda)} = \left[1 + \left(\frac{x_3^{\lambda+1} + x_4^{\lambda+1}}{x_1^{\lambda+1} + x_2^{\lambda+1}}\right)^{1/(\lambda+1)}\right]^{-1}. \tag{A1.8}$$

The maximum likelihood and minimum chi-squared estimators follow immediately from (A1.8) as

$$\hat{\theta}^{(0)} = (x_1 + x_2)/n$$

and

$$\hat{\theta}^{(1)} = \left[1 + \left(\frac{x_3^2 + x_4^2}{x_1^2 + x_2^2}\right)^{1/2}\right]^{-1}.$$

In equation (A1.7), substitute $T_1 = (3 - 8\theta + 8\theta^2)/(2\theta^4(1-\theta)^4)$, $T_2 = 2/(\theta^3(1-\theta)^3)$, and $\mu_{20} = 1/(\theta(1-\theta))$. Hence

$$D_{HL}[\hat{\theta}^{(\lambda)}, \hat{\theta}^{(0)}] = \lambda^2(3 - 8\theta + 8\theta^2)/(4\theta(1-\theta)) - 2\lambda.$$

The least deficient value of λ is given by Theorem A1.1 to be $\lambda^* = T_2/T_1 = 4\theta(1-\theta)/(3 - 8\theta + 8\theta^2)$. As θ varies in the parameter space $[0, 1]$, the values of λ^* are given by the curve in Figure A1.1. We see that if the true value of θ lies in the interval $[0.25, 0.75]$, then the optimal value of λ lies in the interval

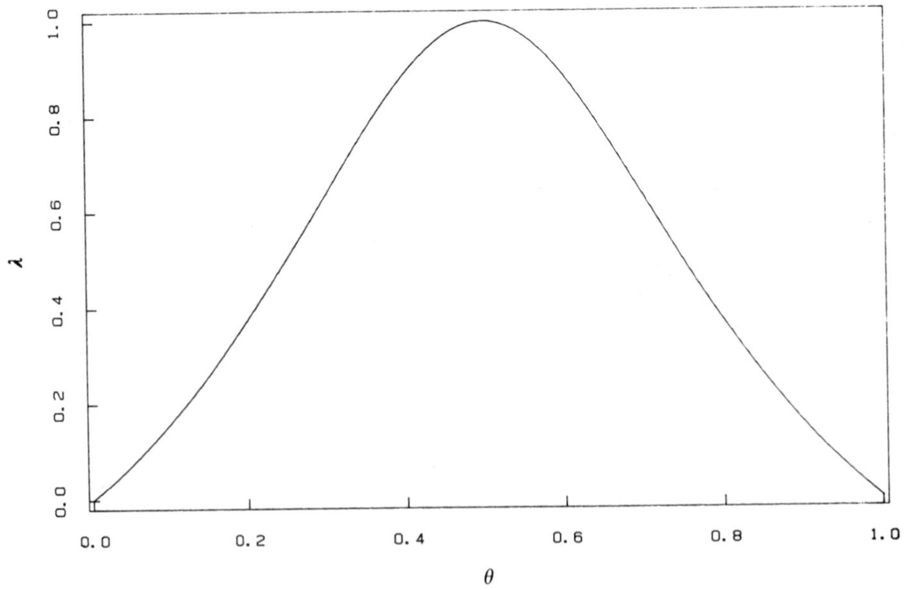

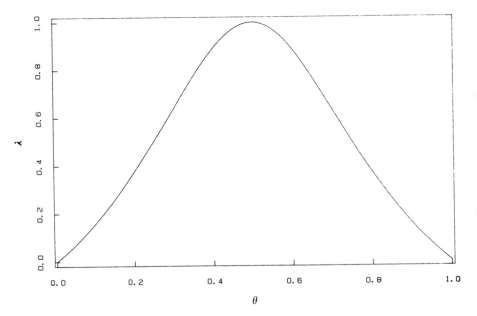

Figure A1.1. Least deficient choice of λ for the minimum power-divergence estimator defined in (A1.8) as a function of θ.

[0.5, 1.0]. Furthermore, the MLE ($\lambda = 0$) is optimal only in the two extreme cases $\theta = 0$ and $\theta = 1$. Since the minimax criterion guarantees the best λ for the worst θ, it should not be surprising that the minimax solution is $\lambda = 0$.

A2. Characterization of the Generalized Minimum Power-Divergence Estimate (Section 3.5)

Consider a fixed probability vector $\mathbf{p}$ of length k, a value $\lambda \in (-\infty, \infty)$, and a set of $s + 1$ constraints on a vector $\mathbf{m}$ (also of length k and whose elements are nonnegative)

$$\sum_{i=1}^{k} c_{ij} m_i = n\theta_j; \qquad j = 0, 1, \ldots, s, \tag{A2.1}$$

with $c_{i0} = 1$, for $i = 1, \ldots, k$ and $\theta_0 = 1$ (i.e., $\sum_{i=1}^{k} m_i = n$). If there exists a vector $\mathbf{m}^{*(\lambda)}$ satisfying (A2.1) that can be written in the form

$$\frac{1}{\lambda}\left[\left(\frac{m_i^{*(\lambda)}}{np_i}\right)^{\lambda} - 1\right] = \sum_{j=0}^{s} c_{ij}\tau_j \tag{A2.2}$$

(where the case $\lambda = 0$ is interpreted as the limit as $\lambda \to 0$), then

$$2I^{\lambda}(\mathbf{m} : n\mathbf{p}) \geq 2I^{\lambda}(\mathbf{m}^{*(\lambda)} : n\mathbf{p}), \tag{A2.3}$$

for all $\mathbf{m}$ satisfying (A2.1), and $\mathbf{m}^{*(\lambda)}$ is unique. The proof of this result is split into two cases

$\lambda \neq -1$

Define $\psi(x) = (x^{-\lambda} - 1)/\lambda(\lambda + 1)$ where the case $\lambda = 0$ is defined by the limit as $\lambda \to 0$. Since $\psi(x)$ is strictly convex, it follows that

$$\psi(y) - \psi(x) > (y - x)\psi'(x); \qquad \text{for all } y \neq x.$$

Set $y = np_i/m_i$ and $x = np_i/m_i^{*(\lambda)}$, then for $m_i \neq m_i^{*(\lambda)}$

$$\frac{1}{\lambda(\lambda + 1)}\left[\left(\frac{m_i}{np_i}\right)^{\lambda} - \left(\frac{m_i^{*(\lambda)}}{np_i}\right)^{\lambda}\right] > \left(\frac{np_i}{m_i^{*(\lambda)}} - \frac{np_i}{m_i}\right)\frac{1}{\lambda + 1}\left(\frac{m_i^{*(\lambda)}}{np_i}\right)^{\lambda+1}$$

(again $\lambda = 0$ is interpreted as the limit as $\lambda \to 0$). Weighting by m_i and summing over $i = 1, \ldots, k$ gives

$$\frac{1}{\lambda(\lambda + 1)}\left[\sum_{i=1}^{k} m_i\left(\frac{m_i}{np_i}\right)^{\lambda} - \sum_{i=1}^{k} m_i\left(\frac{m_i^{*(\lambda)}}{np_i}\right)^{\lambda}\right] > \frac{1}{\lambda + 1}\sum_{i=1}^{k} (m_i - m_i^{*(\lambda)})\left(\frac{m_i^{*(\lambda)}}{np_i}\right)^{\lambda}. \tag{A2.4}$$

Applying equations (A2.1) and (A2.2) shows

$$\frac{1}{\lambda}\sum_{i=1}^{k} m_i\left[\left(\frac{m_i^{*(\lambda)}}{np_i}\right)^{\lambda} - 1\right] = \sum_{i=1}^{k}\sum_{j=0}^{s} m_i c_{ij}\tau_j$$
$$= \sum_{j=0}^{s} n\theta_j\tau_j$$
$$= \sum_{i=1}^{k}\sum_{j=0}^{s} m_i^{*(\lambda)} c_{ij}\tau_j$$
$$= \frac{1}{\lambda}\sum_{i=1}^{k} m_i^{*(\lambda)}\left[\left(\frac{m_i^{*(\lambda)}}{np_i}\right)^{\lambda} - 1\right], \tag{A2.5}$$

and therefore for $\lambda \neq 0$,

$$\sum_{i=1}^{k} m_i\left(\frac{m_i^{*(\lambda)}}{np_i}\right)^{\lambda} = \sum_{i=1}^{k} m_i^{*(\lambda)}\left(\frac{m_i^{*(\lambda)}}{np_i}\right)^{\lambda}. \tag{A2.6}$$

Using (A2.5) on the left-hand side of (A2.4), and (A2.6) (when $\lambda \neq 0$) on the right-hand side of (A2.4), we obtain the result (A2.3).

$\lambda = -1$

Define $\psi(x) = \log(x^{-1})$, which is strictly convex, so

$$\log(y^{-1}) - \log(x^{-1}) > (x - y)/x; \qquad \text{for all } x \neq y.$$

Set $y = m_i/np_i$ and $x = m_i^{*(-1)}/np_i$, then for $m_i \neq m_i^{*(-1)}$

$$\log\left(\frac{np_i}{m_i}\right) - \log\left(\frac{np_i}{m_i^{*(-1)}}\right) > \left(\frac{m_i^{*(-1)}}{np_i} + \frac{m_i}{np_i}\right)\left(\frac{np_i}{m_i^{*(-1)}}\right).$$

Weighting by np_i and summing over $i = 1, \ldots, k$ gives

$$\sum_{i=1}^{k} np_i\log\left(\frac{np_i}{m_i}\right) - \sum_{i=1}^{k} np_i\log\left(\frac{np_i}{m_i^{*(-1)}}\right) > \sum_{i=1}^{k}(m_i^{*(-1)} - m_i)\left(\frac{np_i}{m_i^{*(-1)}}\right). \tag{A2.7}$$

Now using (A2.6) (which is easily seen to be true for $\lambda = -1$) on the right-hand side of (A2.7) yields the result (A2.3).

The uniqueness of $\mathbf{m}^{*(\lambda)}$ for each $\lambda \in (-\infty, \infty)$ follows from the strict convexity of $\psi(x)$ for both cases $\lambda = -1$ and $\lambda \neq -1$.

A3. Characterization of the Lancaster-Additive Model (Section 3.5)

The Lancaster-additive model for the three-dimensional contingency table can be defined from (3.50) as

$$\frac{\pi_{ijl}}{\pi_{i++}\pi_{+j+}\pi_{++l}} = \alpha_{ij} + \beta_{il} + \gamma_{jl}. \tag{A3.1}$$

To show how this model can be derived from the generalized minimum power-divergence estimate given by (3.44) and (3.45) (with $\lambda = 1$), we need to reexpress the elements p_i and $m_i = n\pi_i$ of vectors $\mathbf{p}$ and $\mathbf{m} = n\boldsymbol{\pi}$ as three-dimensional matrix elements p_{ijl} and $m_{ijl} = n\pi_{ijl}$. First, consider the equations (3.44) required to fix the two-dimensional margin m_{ij+}; these constraints can be expressed as

$$\sum_{i',j',l'} c^{(12)}_{i'j'l'(ij)} m_{i'j'l'} = n\theta^{(12)}_{ij}; \qquad \text{for each } i, j \tag{A3.2}$$

(where the summation is over all possible values of the indices i', j', and l'), $c^{(12)}_{i'j'l'(ij)} = 1$ if both $i' = i$ and $j' = j$, and 0 otherwise. Equation (A3.2) collapses to $m_{ij+} = n\theta^{(12)}_{ij}$ where $\theta^{(12)}_{ij}$ is fixed for each i, j. Two similar sets of equations fix the margins m_{i+l} and m_{+jl}, for suitably defined coefficients $c^{(13)}_{i'j'l'(il)}$ and $c^{(23)}_{i'j'l'(jl)}$. Finally, we include the constraint $\sum_{i',j',l'} m_{i'j'l'} = n$. Now set $\lambda = 1$ and $p_{ijl} = \pi_{i++}\pi_{+j+}\pi_{++l}$ in (3.45) giving

$$\begin{aligned}\frac{\pi^{*(1)}_{i'j'l'}}{\pi_{i'++}\pi_{+j'+}\pi_{++l'}} - 1 &= \tau_0 + \sum_{i,j} c^{(12)}_{i'j'l'(ij)}\tau^{(12)}_{ij} + \sum_{i,l} c^{(13)}_{i'j'l'(il)}\tau^{(13)}_{il} + \sum_{j,l} c^{(23)}_{i'j'l'(jl)}\tau^{(23)}_{jl}\\ &= \tau_0 + \tau^{(12)}_{i'j'} + \tau^{(13)}_{i'l'} + \tau^{(23)}_{j'l'},\end{aligned}$$

which can be expressed in the general form (A3.1).

A4. Proof of Results (i), (ii), and (iii) (Section 4.1)

The results (i), (ii), and (iii) in Section 4.1 are fundamental building blocks for deriving the asymptotic chi-squared distribution of the power-divergence statistic as $n \to \infty$. Because of the importance of these foundations, we shall provide the background to their derivation. Further details are available in, for example, Rao (1973, chapter 6), Bishop et al. (1975, chapter 14), and Read (1982).

Result (i). *Assume* $\mathbf{X}$ *is a random row vector with a multinomial distribution* $\mathrm{Mult}_k(n, \boldsymbol{\pi})$, *and* $\boldsymbol{\pi} = \boldsymbol{\pi}_0$ *from* (4.2). *Then* $\mathbf{W}_n = \sqrt{n}(\mathbf{X}/n - \boldsymbol{\pi}_0)$ *converges in distribution to a multivariate normal random vector* $\mathbf{W}$ *as* $n \to \infty$. *The mean vector and covariance matrix of* $\mathbf{W}_n$ *(and* $\mathbf{W}$*) are*

$$\begin{aligned} E(\mathbf{W}_n) &= \mathbf{0},\\ \operatorname{cov}(\mathbf{W}_n) &= D_{\boldsymbol{\pi}_0} - \boldsymbol{\pi}_0'\boldsymbol{\pi}_0, \end{aligned} \tag{A4.1}$$

where $D_{\boldsymbol{\pi}_0}$ *is the* $k \times k$ *diagonal matrix based on* $\boldsymbol{\pi}_0$.

Proof. The mean and covariance of $\mathbf{W}_n$ in (A4.1) are derived immediately from

$$\begin{aligned} E(X_i) &= n\pi_{0i},\\ \operatorname{var}(X_i) &= n\pi_{0i}(1 - \pi_{0i}), \end{aligned}$$

and

$$\text{cov}(X_i, X_j) = -n\pi_{0i}\pi_{0j};$$

therefore $E(\mathbf{X}) = n\boldsymbol{\pi}_0$ and $\text{cov}(\mathbf{X}) = n(D_{\boldsymbol{\pi}_0} - \boldsymbol{\pi}_0'\boldsymbol{\pi}_0)$.

The asymptotic normality of $\mathbf{W}_n$ follows by showing that the moment-generating function (MGF) of $\mathbf{W}_n$ converges to the MGF of a multivariate normal random vector $\mathbf{W}$ with mean and covariance given by (A4.1). Specifically, the MGF of $\mathbf{W}_n$ is

$$\begin{aligned} M_{\mathbf{W}_n}(\mathbf{v}) &= E[\exp(\mathbf{v}\mathbf{W}_n')] \\ &= \exp(-n^{1/2}\mathbf{v}\boldsymbol{\pi}_0')E[\exp(n^{-1/2}\mathbf{v}\mathbf{X}')] \\ &= \exp(-n^{1/2}\mathbf{v}\boldsymbol{\pi}_0')M_{\mathbf{X}}(n^{-1/2}\mathbf{v}), \end{aligned}$$

where $M_{\mathbf{X}}(\mathbf{v}) = [\sum_{i=1}^k \pi_{0i}\exp(v_i)]^n$ is the MGF of the multinomial random vector $\mathbf{X}$. Therefore

$$M_{\mathbf{W}_n}(\mathbf{v}) = \left[\sum_{i=1}^k \pi_{0i}\exp(n^{-1/2}(v_i - \mathbf{v}\boldsymbol{\pi}_0'))\right]^n. \tag{A4.2}$$

Expanding (A4.2) in a Taylor series gives

$$\begin{aligned} M_{\mathbf{W}_n}(\mathbf{v}) &= \left[1 + \frac{1}{\sqrt{n}}\sum_{i=1}^k \pi_{0i}(v_i - \mathbf{v}\boldsymbol{\pi}_0') + \frac{1}{2n}\sum_{i=1}^k \pi_{0i}(v_i - \mathbf{v}\boldsymbol{\pi}_0')^2 + o(n^{-1})\right]^n \\ &= [1 + \mathbf{v}(D_{\boldsymbol{\pi}_0} - \boldsymbol{\pi}_0'\boldsymbol{\pi}_0)\mathbf{v}'/2n + o(n^{-1})]^n \to \exp[\mathbf{v}(D_{\boldsymbol{\pi}_0} - \boldsymbol{\pi}_0'\boldsymbol{\pi}_0)\mathbf{v}'/2], \end{aligned}$$

as $n \to \infty$. This is the MGF of the multivariate normal random vector $\mathbf{W}$ with mean vector $\mathbf{0}$ and covariance matrix $D_{\boldsymbol{\pi}_0} - \boldsymbol{\pi}_0'\boldsymbol{\pi}_0$, which proves the result. □

Result (ii).

$$X^2 = \sum_{i=1}^k \frac{(X_i - n\pi_{0i})^2}{n\pi_{0i}}$$

can be written as a quadratic form in $\mathbf{W}_n = \sqrt{n}(\mathbf{X}/n - \boldsymbol{\pi}_0)$, *and* X^2 *converges in distribution (as* $n \to \infty$*) to the quadratic form of the multivariate normal random vector* $\mathbf{W}$ *in result* (i).

Proof. It is straightforward to show that

$$\begin{aligned} X^2 &= \sum_{i=1}^k \sqrt{n}(X_i/n - \pi_{0i})\frac{1}{\pi_{0i}}\sqrt{n}(X_i/n - \pi_{0i}) \\ &= \mathbf{W}_n(D_{\boldsymbol{\pi}_0})^{-1}\mathbf{W}_n'. \end{aligned}$$

From result (i) we know that $\mathbf{W}_n$ converges in distribution to $\mathbf{W}$, which is a multivariate normal random vector with mean and covariance given by (A4.1). We now appeal to the general result that for any continuous function $g(\cdot)$, $g(\mathbf{W}_n)$ converges in distribution to $g(\mathbf{W})$ (Rao, 1973, p. 124). Consequently,

$\mathbf{W}_n(D_{\boldsymbol{\pi}_0})^{-1}\mathbf{W}'_n$ converges in distribution to $\mathbf{W}(D_{\boldsymbol{\pi}_0})^{-1}\mathbf{W}'$, which proves the result. □

Result (iii). *Certain quadratic forms (indicated in the proof to follow) of multivariate normal random vectors are distributed as chi-squared random variables. In the case of results* (i) *and* (ii), *where* $X^2 = \mathbf{W}_n(D_{\boldsymbol{\pi}_0})^{-1}\mathbf{W}'_n$, X^2 *converges in distribution to a central chi-squared random variable with* $k-1$ *degrees of freedom.*

PROOF. We rely on the following general result given by Bishop et al. (1975, p. 473). Assume $\mathbf{U} = (U_1, U_2, \ldots, U_k)$ has a multivariate normal distribution with mean vector $\mathbf{0}$ and covariance matrix Σ, and $Y = \mathbf{U}B\mathbf{U}'$ for some symmetric matrix B. Then Y has the same distribution as $\sum_{i=1}^{k} \zeta_i Z_i^2$, where the Z_i^2s are independent chi-squared random variables, each with one degree of freedom, and the ζ_is are the eigenvalues of $B^{1/2}\Sigma(B^{1/2})'$.

In the present case

$$\mathbf{U} = \mathbf{W},$$

$$B = (D_{\boldsymbol{\pi}_0})^{-1},$$

$$\Sigma = D_{\boldsymbol{\pi}_0} - \boldsymbol{\pi}'_0\boldsymbol{\pi}_0,$$

and

$$B^{1/2}\Sigma(B^{1/2})' = (D_{\boldsymbol{\pi}_0})^{-1/2}(D_{\boldsymbol{\pi}_0} - \boldsymbol{\pi}'_0\boldsymbol{\pi}_0)(D_{\boldsymbol{\pi}_0})^{-1/2}$$
$$= I - \sqrt{\boldsymbol{\pi}'_0}\sqrt{\boldsymbol{\pi}_0},$$

where I is the $k \times k$ identity matrix, and $\sqrt{\boldsymbol{\pi}_0} = (\sqrt{\pi_{01}}, \sqrt{\pi_{02}}, \ldots, \sqrt{\pi_{0k}})$. It can be shown that $k-1$ of the eigenvalues of $I - \sqrt{\boldsymbol{\pi}'_0}\sqrt{\boldsymbol{\pi}_0}$ equal 1, and the remaining eigenvalue equals 0. Consequently the distribution of $\mathbf{W}(D_{\boldsymbol{\pi}_0})^{-1}\mathbf{W}'$ is the same as that of $\sum_{i=1}^{k-1} Z_i^2$, which is chi-squared with $k-1$ degrees of freedom (since the Z_is are independent). From result (ii), this is also the asymptotic distribution of $X^2 = \mathbf{W}_n(D_{\boldsymbol{\pi}_0})^{-1}\mathbf{W}'_n$. □

A5. Statement of Birch's Regularity Conditions and Proof that the Minimum Power-Divergence Estimator Is BAN (Section 4.1)

Throughout this section we assume $\mathbf{X}$ is a multinomial $\text{Mult}_k(n, \boldsymbol{\pi})$ random row vector, and the hypotheses

$$H_0: \boldsymbol{\pi} \in \Pi_0$$

versus (A5.1)

$$H_1: \boldsymbol{\pi} \notin \Pi_0$$

can be reparameterized by assuming that under H_0, the (unknown) vector of true probabilities $\boldsymbol{\pi}^* = (\pi_1^*, \pi_2^*, \ldots, \pi_k^*) \in \Pi_0$ is a function of parameters $\boldsymbol{\theta}^* = (\theta_1^*, \theta_2^*, \ldots, \theta_s^*) \in \Theta_0$ where $s < k - 1$. More specifically, we define a function $\mathbf{f}(\boldsymbol{\theta})$ that maps each element of the subset $\Theta_0 \subset R^s$ into the subset $\Pi_0 \subset \Delta_k = \{\mathbf{p} = (p_1, p_2, \ldots, p_k): p_i \geq 0; i = 1, \ldots, k \text{ and } \sum_{i=1}^k p_i = 1\}$. Therefore (A5.1) can be reparameterized in terms of the pair $(\mathbf{f}, \Theta_0)$ as

$$H_0\text{: There exists a } \boldsymbol{\theta}^* \in \Theta_0 \text{ such that } \boldsymbol{\pi} = \mathbf{f}(\boldsymbol{\theta}^*)\ (\equiv \boldsymbol{\pi}^*)$$

versus (A5.2)

$$H_1\text{: No } \boldsymbol{\theta}^* \text{ exists in } \Theta_0 \text{ for which } \boldsymbol{\pi} = \mathbf{f}(\boldsymbol{\theta}^*).$$

Instead of describing the estimation of $\boldsymbol{\pi}^*$ in terms of choosing a value $\hat{\boldsymbol{\pi}} \in \Pi_0$ that minimizes a specific objective function (e.g., minimum power-divergence estimation in (4.7)), we can consider choosing $\hat{\boldsymbol{\theta}} \in \bar{\Theta}_0$ (where $\bar{\Theta}_0$ represents the closure of Θ_0) for which $\mathbf{f}(\hat{\boldsymbol{\theta}})$ minimizes the same objective function, and then set $\hat{\boldsymbol{\pi}} = \mathbf{f}(\hat{\boldsymbol{\theta}})$. This reparameterization provides a simpler framework within which to describe the properties of the minimum power-divergence estimator $\hat{\boldsymbol{\pi}}^{(\lambda)} = \mathbf{f}(\hat{\boldsymbol{\theta}}^{(\lambda)})$ of $\boldsymbol{\pi}^*$, or $\hat{\boldsymbol{\theta}}^{(\lambda)}$ of $\boldsymbol{\theta}^*$ defined by

$$I^\lambda(\mathbf{X}/n : \mathbf{f}(\hat{\boldsymbol{\theta}}^{(\lambda)})) = \inf_{\boldsymbol{\theta} \in \Theta_0} I^\lambda(\mathbf{X}/n : \mathbf{f}(\boldsymbol{\theta})). \tag{A5.3}$$

In order to ensure that the minimum power-divergence estimator $\hat{\boldsymbol{\theta}}^{(\lambda)}$ exists and converges in probability to $\boldsymbol{\theta}^*$ as $n \to \infty$, it is necessary to specify some regularity conditions (Birch, 1964) on $\mathbf{f}$ and Θ_0 under H_0. These conditions ensure that the null model really has s parameters, and that $\mathbf{f}$ satisfies various "smoothness" requirements discussed more fully by Bishop et al. (1975, pp. 509–511). Assuming H_0 (i.e., there exists a $\boldsymbol{\theta}^* \in \Theta_0$ such that $\boldsymbol{\pi} = \mathbf{f}(\boldsymbol{\theta}^*)\ (\equiv \boldsymbol{\pi}^*)$) and that $s < k - 1$, Birch's regularity conditions are:

(a) $\boldsymbol{\theta}^*$ is an interior point of Θ_0, and there is an s-dimensional neighborhood of $\boldsymbol{\theta}^*$ completely contained in Θ_0;
(b) $\pi_i^* \equiv f_i(\boldsymbol{\theta}^*) > 0$ for $i = 1, \ldots, k$. Hence $\boldsymbol{\pi}^*$ is an interior point of the $(k-1)$-dimensional simplex Δ_k;
(c) The mapping $\mathbf{f}: \Theta_0 \to \Delta_k$ is totally differentiable at $\boldsymbol{\theta}^*$, so that the partial derivatives of f_i with respect to each θ_j exist at $\boldsymbol{\theta}^*$ and $\mathbf{f}(\boldsymbol{\theta})$ has a linear approximation at $\boldsymbol{\theta}^*$ given by
$$\mathbf{f}(\boldsymbol{\theta}) = \mathbf{f}(\boldsymbol{\theta}^*) + (\boldsymbol{\theta} - \boldsymbol{\theta}^*)(\partial \mathbf{f}(\boldsymbol{\theta}^*)/\partial \boldsymbol{\theta})' + o(|\boldsymbol{\theta} - \boldsymbol{\theta}^*|) \text{ as } \boldsymbol{\theta} \to \boldsymbol{\theta}^*,$$
where $\partial \mathbf{f}(\boldsymbol{\theta}^*)/\partial \boldsymbol{\theta}$ is a $k \times s$ matrix with (i,j)th element $\partial f_i(\boldsymbol{\theta}^*)/\partial \theta_j$;
(d) The Jacobian matrix $\partial \mathbf{f}(\boldsymbol{\theta}^*)/\partial \boldsymbol{\theta}$ is of full rank s;
(e) The inverse mapping $\mathbf{f}^{-1}: \Pi_0 \to \Theta_0$ is continuous at $\mathbf{f}(\boldsymbol{\theta}^*) = \boldsymbol{\pi}^*$; and
(f) The mapping $\mathbf{f}: \Theta_0 \to \Delta_k$ is continuous at every point $\boldsymbol{\theta} \in \Theta_0$.

These regularity conditions are necessary in order to establish the key asymptotic expansion of the power-divergence estimator $\hat{\boldsymbol{\theta}}^{(\lambda)}$ of $\boldsymbol{\theta}^*$ under H_0,

$$\hat{\boldsymbol{\theta}}^{(\lambda)} = \boldsymbol{\theta}^* + (\mathbf{X}/n - \boldsymbol{\pi}^*)(D_{\boldsymbol{\pi}^*})^{-1/2} A (A'A)^{-1} + o_p(n^{-1/2}), \tag{A5.4}$$

where $D_{\boldsymbol{\pi}^*}$ is the $k \times k$ diagonal matrix based on $\boldsymbol{\pi}^*$, and A is the $k \times s$ matrix with (i,j)th element $(\pi_i^*)^{-1/2}\partial f_i(\boldsymbol{\theta}^*)/\partial\theta_j$. An estimator that satisfies (A5.4) is called best asymptotically normal (BAN). This expansion plays a central role in deriving the asymptotic distribution of the power-divergence statistic under H_0 (Section A6). The rest of this section is devoted to proving (A5.4). The asymptotic normality of $\hat{\boldsymbol{\theta}}^{(\lambda)}$ is derived as a corollary to Theorem A5.1.

Theorem A5.1. *Assume H_0 in* (A5.2) *holds together with the regularity conditions* (a)–(f). *Let*

$$\bar{\Theta}_0 = \begin{cases} \text{the closure of } \Theta_0, \text{ if } \Theta_0 \text{ is bounded,} \\ \text{the closure of } \Theta_0 \text{ and a point at infinity, otherwise.} \end{cases}$$

If λ is fixed and $\hat{\boldsymbol{\theta}}^{(\lambda)}$ is any value of $\boldsymbol{\theta} \in \bar{\Theta}_0$ for which (A5.3) *holds, then $\hat{\boldsymbol{\theta}}^{(\lambda)}$ is BAN. That is,*

$$\hat{\boldsymbol{\theta}}^{(\lambda)} = \boldsymbol{\theta}^* + (\mathbf{X}/n - \boldsymbol{\pi}^*)(D_{\boldsymbol{\pi}^*})^{-1/2}A(A'A)^{-1} + o_p(n^{-1/2})$$

as $n \to \infty$, from (A5.4).

Remark. This theorem is a direct generalization of that provided by Birch (1964) for the maximum likelihood estimator ($\lambda = 0$) and appeared in Read (1982). However the assumption that $\mathbf{f}$ is continuous (regularity condition (**f**)) simplifies our statement of the theorem, because the existence of such a $\hat{\boldsymbol{\theta}}^{(\lambda)} \in \bar{\Theta}_0$ is ensured. Recently Cox (1984) presented a more elementary proof of a version of Birch's theorem using the implicit value theorem. Cox's regularity conditions are slightly stronger than those of Birch, but yield a stronger conclusion.

To prove the theorem we shall follow closely Birch's proof by presenting a series of lemmas. The details of proving each lemma will be omitted whenever we can reference Birch. Throughout the proof, the notation $|\ |$ defines the Euclidean norm (i.e., $|\mathbf{p}| = (\sum_{i=1}^k p_i^2)^{1/2}$).

Lemma A5.1. *If $\mathbf{p} \in \Delta_k$ with $p_i > 0$ and $\boldsymbol{\theta} \in \Theta_0$ with $f_i(\boldsymbol{\theta}) > 0$; $i = 1, \ldots, k$, then*

$$2I^\lambda(\mathbf{p} : \mathbf{f}(\boldsymbol{\theta})) = \sum_{i=1}^k \gamma_i (p_i - f_i(\boldsymbol{\theta}))^2 \tag{A5.5}$$

where $\gamma_i = \alpha_i^{\lambda-1}/f_i(\boldsymbol{\theta})^\lambda$ for some α_i between p_i and $f_i(\boldsymbol{\theta})$; $i = 1, \ldots, k$. Furthermore (for λ fixed) there exists a $\zeta > 0$ such that

$$2I^\lambda(\mathbf{p} : \mathbf{f}(\boldsymbol{\theta})) \geq \zeta \tag{A5.6}$$

whenever $|\mathbf{p} - \mathbf{f}(\boldsymbol{\theta})| > \delta > 0$.

Proof of Lemma A5.1. Equation (A5.5) follows by expanding the summand of $2I^\lambda(\mathbf{p} : \mathbf{f}(\boldsymbol{\theta}))$ in the Taylor series

$$\frac{2p_i}{\lambda(\lambda+1)}\left[\left(\frac{p_i}{f_i(\boldsymbol{\theta})}\right)^\lambda - 1\right] = \frac{2}{\lambda}\left(\frac{f_i(\boldsymbol{\theta})}{f_i(\boldsymbol{\theta})}\right)^\lambda (p_i - f_i(\boldsymbol{\theta})) + \frac{\alpha_i^{\lambda-1}}{f_i(\boldsymbol{\theta})^\lambda}(p_i - f_i(\boldsymbol{\theta}))^2,$$

where α_i is between p_i and $f_i(\boldsymbol{\theta})$. Summing over i gives (A5.5). Inequality (A5.6) follows from analysis of (A5.5). □

Lemma A5.2. *As* $\mathbf{p} \to \boldsymbol{\pi}^*$ *and* $\boldsymbol{\theta} \to \boldsymbol{\theta}^*$, *then*

$$2I^\lambda(\mathbf{p}:\mathbf{f}(\boldsymbol{\theta})) = |(\mathbf{p} - \boldsymbol{\pi}^*)(D_{\boldsymbol{\pi}^*})^{-1/2} - (\boldsymbol{\theta} - \boldsymbol{\theta}^*)A'|^2 + o(|\mathbf{p} - \boldsymbol{\pi}^*|^2 + |\boldsymbol{\theta} - \boldsymbol{\theta}^*|^2).$$

PROOF OF LEMMA A5.2. This result is obtained by applying equation (A5.5) and following the proof of Birch's lemma 2. □

Lemma A5.3. *If* $\boldsymbol{\theta}^{\#}(\mathbf{p}) = \boldsymbol{\theta}^* + (\mathbf{p} - \boldsymbol{\pi}^*)(D_{\boldsymbol{\pi}^*})^{-1/2}A(A'A)^{-1}$, *then*

$$2I^\lambda(\mathbf{p}:\mathbf{f}(\boldsymbol{\theta})) = |(\mathbf{p} - \boldsymbol{\pi}^*)(D_{\boldsymbol{\pi}^*})^{-1/2} - (\boldsymbol{\theta}^{\#}(\mathbf{p}) - \boldsymbol{\theta}^*)A'|^2 + |(\boldsymbol{\theta}^{\#}(\mathbf{p}) - \boldsymbol{\theta}^*)A'|^2$$
$$+ o(|\mathbf{p} - \boldsymbol{\pi}^*|^2 + |\boldsymbol{\theta}^{\#}(\mathbf{p}) - \boldsymbol{\theta}|^2)$$

as $\boldsymbol{\theta} \to \boldsymbol{\theta}^*$ *and* $\mathbf{p} \to \boldsymbol{\pi}^*$.

PROOF OF LEMMA A5.3. The proof is directly analogous to lemma 3 of Birch (1964). □

Lemma A5.4. *Let* $\boldsymbol{\theta}^{(\lambda)}(\mathbf{p}) \in \bar{\Theta}_0$ *satisfy*

$$2I^\lambda(\mathbf{p}:\mathbf{f}(\boldsymbol{\theta}^{(\lambda)}(\mathbf{p}))) = \inf_{\boldsymbol{\theta} \in \Theta_0} 2I^\lambda(\mathbf{p}:\mathbf{f}(\boldsymbol{\theta})).$$

Then

$$\boldsymbol{\theta}^{(\lambda)}(\mathbf{p}) - \boldsymbol{\theta}^{\#}(\mathbf{p}) = o(|\mathbf{p} - \boldsymbol{\pi}^*|),$$

as $\mathbf{p} \to \boldsymbol{\pi}^*$. *That is,*

$$\boldsymbol{\theta}^{(\lambda)}(\mathbf{p}) = \boldsymbol{\theta}^* + (\mathbf{p} - \boldsymbol{\pi}^*)(D_{\boldsymbol{\pi}^*})^{-1/2}A(A'A)^{-1} + o(|\mathbf{p} - \boldsymbol{\pi}^*|).$$

PROOF OF LEMMA A5.4. This follows by applying Lemmas A5.1 and A5.3 in a directly analogous way to lemma 4 of Birch (1964). □

PROOF OF THEOREM A5.1. Note that $\mathbf{X}/n = \boldsymbol{\pi}^* + O_p(n^{-1/2})$ (which is a direct consequence of the asymptotic normality of $\sqrt{n}(\mathbf{X}/n - \boldsymbol{\pi}^*)$ proved in Section A4). Therefore setting $\mathbf{p} = \mathbf{X}/n$ in Lemma A5.4, and defining $\hat{\boldsymbol{\theta}}^{(\lambda)} = \boldsymbol{\theta}^{(\lambda)}(\mathbf{X}/n)$ gives (A5.4) as required. (For more details regarding substitution of stochastic sequences into deterministic equations, see Bishop et al., 1975, section 14.4.5.) This ends the proof. □

Finally we show that under the conditions of this section, any BAN-estimator $\hat{\boldsymbol{\theta}}$ of $\boldsymbol{\theta}^*$ has an asymptotic normal distribution, which justifies the terminology best asymptotically normal. In particular, the minimum power-divergence estimator $\hat{\boldsymbol{\theta}}^{(\lambda)}$ in asymptotically normal for all $\lambda \in (-\infty, \infty)$.

Corollary A5.1. *Assume* H_0 *and Birch's regularity conditions* (a)–(f). *Then provided* $\hat{\boldsymbol{\theta}}$ *satisfies*

$$\hat{\boldsymbol{\theta}} = \boldsymbol{\theta}^* + (\mathbf{X}/n - \boldsymbol{\pi}^*)(D_{\boldsymbol{\pi}^*})^{-1/2}A(A'A)^{-1} + o_p(n^{-1/2}), \tag{A5.7}$$

the asymptotic distribution of $\sqrt{n}(\hat{\boldsymbol{\theta}} - \boldsymbol{\theta}^*)$ *is multivariate normal with mean vector* $\mathbf{0}$ *and covariance matrix* $(A'A)^{-1}$, *where* A *is the* $k \times s$ *matrix with* (i,j)*th element* $(\pi_i^*)^{-1/2}\partial f_i(\boldsymbol{\theta}^*)/\partial\theta_j$.

Proof. From (A5.7), we have

$$\sqrt{n}(\hat{\boldsymbol{\theta}} - \boldsymbol{\theta}^*) = \sqrt{n}(\mathbf{X}/n - \boldsymbol{\pi}^*)(D_{\boldsymbol{\pi}^*})^{-1/2}A(A'A)^{-1} + o_p(1).$$

From result (i) of Section A4, we know that $\sqrt{n}(\mathbf{X}/n - \boldsymbol{\pi}^*)$ has an asymptotic normal distribution with mean vector $\mathbf{0}$ and covariance matrix $D_{\boldsymbol{\pi}^*} - \boldsymbol{\pi}^{*\prime}\boldsymbol{\pi}^*$. Therefore $\sqrt{n}(\mathbf{X}/n - \boldsymbol{\pi}^*)(D_{\boldsymbol{\pi}^*})^{-1/2}A(A'A)^{-1}$ is also asymptotically normally distributed with mean vector

$$E[\sqrt{n}(\mathbf{X}/n - \boldsymbol{\pi}^*)](D_{\boldsymbol{\pi}^*})^{-1/2}A(A'A)^{-1} = \mathbf{0},$$

and covariance matrix

$$\begin{aligned}
&\operatorname{cov}[\sqrt{n}(\mathbf{X}/n - \boldsymbol{\pi}^*)(D_{\boldsymbol{\pi}^*})^{-1/2}A(A'A)^{-1}]\\
&\quad = (A'A)^{-1}A'(D_{\boldsymbol{\pi}^*})^{-1/2}(D_{\boldsymbol{\pi}^*} - \boldsymbol{\pi}^{*\prime}\boldsymbol{\pi}^*)(D_{\boldsymbol{\pi}^*})^{-1/2}A(A'A)^{-1}\\
&\quad = (A'A)^{-1}A'(I - \sqrt{\boldsymbol{\pi}^{*\prime}}\sqrt{\boldsymbol{\pi}^*})A(A'A)^{-1}\\
&\quad = (A'A)^{-1},
\end{aligned}$$

since $\sqrt{\boldsymbol{\pi}^*}A = \mathbf{0}$ (where $\sqrt{\boldsymbol{\pi}^*} = (\sqrt{\pi_1^*}, \sqrt{\pi_2^*}, \ldots, \sqrt{\pi_k^*})$). □

A6. Proof of Results (i*), (ii*), and (iii*) (Section 4.1)

Results (i*), (ii*), (iii*) are generalizations of (i), (ii), and (iii) discussed in Section A4. They are fundamental to deriving the asymptotic chi-squared distribution of the power-divergence statistic when s parameters are estimated using BAN estimators (Section A5). Further details of these results can be found in Bishop et al. (1975, chapter 14) and Read (1982).

Result (i*). *Assume* $\mathbf{X}$ *is a random row vector with a multinomial distribution* $\mathrm{Mult}_k(n, \boldsymbol{\pi})$ *and* $\boldsymbol{\pi} = \mathbf{f}(\boldsymbol{\theta}^*) \in \Pi_0$, *from* (A5.2), *for some* $\boldsymbol{\theta}^* = (\theta_1^*, \theta_2^*, \ldots, \theta_s^*) \in \Theta_0 \subset R^s$. *Provided* $\mathbf{f}$ *satisfies Birch's regularity conditions* (a)–(f) *and* $\hat{\boldsymbol{\pi}} \in \Pi_0$ *is a BAN estimator of* $\boldsymbol{\pi}^* \equiv \mathbf{f}(\boldsymbol{\theta}^*)$, *then* $\mathbf{W}_n^* = \sqrt{n}(\mathbf{X}/n - \hat{\boldsymbol{\pi}})$ *converges in distribution to a multivariate normal random vector* $\mathbf{W}^*$ *as* $n \to \infty$. *The mean vector and covariance matrix of* $\mathbf{W}^*$ *are*

$$\begin{aligned}
E(\mathbf{W}^*) &= \mathbf{0}\\
\operatorname{cov}(\mathbf{W}^*) &= D_{\boldsymbol{\pi}^*} - \boldsymbol{\pi}^{*\prime}\boldsymbol{\pi}^* - (D_{\boldsymbol{\pi}^*})^{1/2}A(A'A)^{-1}A'(D_{\boldsymbol{\pi}^*})^{1/2},
\end{aligned} \tag{A6.1}$$

where $D_{\boldsymbol{\pi}^*}$ *is the diagonal matrix based on* $\boldsymbol{\pi}^*$ *and* A *is the* $k \times s$ *matrix with* (i,j)*th element* $(\pi_i^*)^{-1/2}\partial f_i(\boldsymbol{\theta}^*)/\partial\theta_j$.

PROOF. Since $\hat{\boldsymbol{\theta}}$ is BAN, it follows from Section A5 that

$$\hat{\boldsymbol{\theta}} = \boldsymbol{\theta}^* + (\mathbf{X}/n - \boldsymbol{\pi}^*)(D_{\boldsymbol{\pi}^*})^{-1/2}A(A'A)^{-1} + o_p(n^{-1/2}).$$

Therefore Birch's regularity condition (c) (Section A5) gives

$$\mathbf{f}(\hat{\boldsymbol{\theta}}) - \mathbf{f}(\boldsymbol{\theta}^*) = (\hat{\boldsymbol{\theta}} - \boldsymbol{\theta}^*)(\partial \mathbf{f}(\boldsymbol{\theta}^*)/\partial \boldsymbol{\theta})' + o_p(n^{-1/2}),$$

since $\hat{\boldsymbol{\theta}} - \boldsymbol{\theta}^* = O_p(n^{-1/2})$ from Corollary A5.1. Consequently

$$\mathbf{f}(\hat{\boldsymbol{\theta}}) - \mathbf{f}(\boldsymbol{\theta}^*) = (\mathbf{X}/n - \boldsymbol{\pi}^*)L + o_p(n^{-1/2}),$$

where $L = (D_{\boldsymbol{\pi}^*})^{-1/2}A(A'A)^{-1}A'(D_{\boldsymbol{\pi}^*})^{1/2}$, and we can write

$$(\mathbf{X}/n - \boldsymbol{\pi}^*, \hat{\boldsymbol{\pi}} - \boldsymbol{\pi}^*) = (\mathbf{X}/n - \boldsymbol{\pi}^*)(I, L) + o_p(n^{-1/2}), \tag{A6.2}$$

where I is the $k \times k$ identity matrix.

From Section A4 we know that $\sqrt{n}(\mathbf{X}/n - \boldsymbol{\pi}^*)$ has an asymptotic normal distribution with mean vector $\mathbf{0}$ and covariance matrix $D_{\boldsymbol{\pi}^*} - \boldsymbol{\pi}^{*\prime}\boldsymbol{\pi}^*$. Therefore (A6.2) indicates that $\sqrt{n}((\mathbf{X}/n, \hat{\boldsymbol{\pi}}) - (\boldsymbol{\pi}^*, \boldsymbol{\pi}^*))$ will have an asymptotic normal distribution with mean vector $\mathbf{0}$ and covariance matrix $(I, L)'(D_{\boldsymbol{\pi}^*} - \boldsymbol{\pi}^{*\prime}\boldsymbol{\pi}^*)(I, L)$. Hence the joint asymptotic covariance matrix of $\mathbf{X}/n$ and $\hat{\boldsymbol{\pi}}$ can be partitioned as

$$\begin{pmatrix} D_{\boldsymbol{\pi}^*} - \boldsymbol{\pi}^{*\prime}\boldsymbol{\pi}^* & (D_{\boldsymbol{\pi}^*} - \boldsymbol{\pi}^{*\prime}\boldsymbol{\pi}^*)L \\ L'(D_{\boldsymbol{\pi}^*} - \boldsymbol{\pi}^{*\prime}\boldsymbol{\pi}^*) & L'(D_{\boldsymbol{\pi}^*} - \boldsymbol{\pi}^{*\prime}\boldsymbol{\pi}^*)L \end{pmatrix}.$$

Finally we conclude that the difference of the two jointly normal random vectors $\sqrt{n}(\mathbf{X}/n - \boldsymbol{\pi}^*) - \sqrt{n}(\hat{\boldsymbol{\pi}} - \boldsymbol{\pi}^*) = \sqrt{n}(\mathbf{X}/n - \hat{\boldsymbol{\pi}})$ also will be normally distributed, here with mean vector $\mathbf{0}$ and covariance matrix

$$D_{\boldsymbol{\pi}^*} - \boldsymbol{\pi}^{*\prime}\boldsymbol{\pi}^* - (D_{\boldsymbol{\pi}^*} - \boldsymbol{\pi}^{*\prime}\boldsymbol{\pi}^*)L - L'(D_{\boldsymbol{\pi}^*} - \boldsymbol{\pi}^{*\prime}\boldsymbol{\pi}^*) + L'(D_{\boldsymbol{\pi}^*} - \boldsymbol{\pi}^{*\prime}\boldsymbol{\pi}^*)L,$$

which equals (A6.1) by noting that $\boldsymbol{\pi}^*L = \sqrt{\boldsymbol{\pi}^*}A(A'A)^{-1}A'(D_{\boldsymbol{\pi}^*})^{-1/2} = \mathbf{0}$ (where $\sqrt{\boldsymbol{\pi}^*} = (\sqrt{\pi_1^*}, \sqrt{\pi_2^*}, \ldots, \sqrt{\pi_k^*})$), and $L'D_{\boldsymbol{\pi}^*} = (D_{\boldsymbol{\pi}^*})L = (D_{\boldsymbol{\pi}^*})^{1/2}A(A'A)^{-1}A'(D_{\boldsymbol{\pi}^*})^{1/2} = L'(D_{\boldsymbol{\pi}^*})L$. □

Result (ii*).

$$X^2 = \sum_{i=1}^{k} \frac{(X_i - n\hat{\pi}_i)^2}{n\hat{\pi}_i}$$

can be written as a quadratic form in $\mathbf{W}_n^* = \sqrt{n}(\mathbf{X}/n - \hat{\boldsymbol{\pi}})$, *and* X^2 *converges in distribution (as* $n \to \infty$*) to the quadratic form of the multivariate normal random vector* $\mathbf{W}^*$ *in result* (i*).

PROOF. It is straightforward to show that

$$\begin{aligned} X^2 &= \sum_{i=1}^{k} \sqrt{n}(X_i/n - \hat{\pi}_i)\frac{1}{\hat{\pi}_i}\sqrt{n}(X_i/n - \hat{\pi}_i) \\ &= \mathbf{W}_n^*((D_{\boldsymbol{\pi}^*})^{-1} + o_p(1))\mathbf{W}_n^{*\prime}, \end{aligned}$$

since $\hat{\boldsymbol{\pi}} = \boldsymbol{\pi}^* + o_p(1)$ follows from $\hat{\boldsymbol{\pi}}$ being BAN. From result (i*) we know that $\mathbf{W}_n^*$ converges in distribution to $\mathbf{W}^*$, which is a multivariate normal random vector with mean and covariance given by (A6.1). Consequently $\mathbf{W}_n^*(D_{\boldsymbol{\pi}^*})^{-1}\mathbf{W}_n^{*\prime}$ converges in distribution to $\mathbf{W}^*(D_{\boldsymbol{\pi}^*})^{-1}\mathbf{W}^{*\prime}$ as in result (ii) (Section A4). □

Result (iii*). *Certain quadratic forms (indicated in the proof to follow) of multivariate normal random vectors are distributed as chi-squared random variables. In the case of results* (i*) *and* (ii*), *where* $X^2 = \mathbf{W}_n^*(D_{\boldsymbol{\pi}^*})^{-1}\mathbf{W}_n^{*\prime} + o_p(1)$, X^2 *converges in distribution to a central chi-squared random variable with* $k - s - 1$ *degrees of freedom.*

PROOF. This result follows in the same way as result (iii) in Section A4. The quadratic form $\mathbf{W}^*(D_{\boldsymbol{\pi}^*})^{-1}\mathbf{W}^{*\prime}$ has the same distribution as $\sum_{i=1}^k \zeta_i Z_i^2$ where the Z_i^2s are independent chi-squared random variables, each with one degree of freedom, and the ζ_is are the eigenvalues of

$$\begin{aligned}(D_{\boldsymbol{\pi}^*})^{-1/2}&(D_{\boldsymbol{\pi}^*} - \boldsymbol{\pi}^{*\prime}\boldsymbol{\pi}^* - (D_{\boldsymbol{\pi}^*})^{1/2}A(A'A)^{-1}A'(D_{\boldsymbol{\pi}^*})^{1/2})(D_{\boldsymbol{\pi}^*})^{-1/2}\\ &= I - \sqrt{\boldsymbol{\pi}^{*\prime}}\sqrt{\boldsymbol{\pi}^*} - A(A'A)^{-1}A'. \qquad \text{(A6.3)}\end{aligned}$$

It can be shown that $k - s - 1$ of the eigenvalues of (A6.3) equal 1, and the remaining $s + 1$ equal 0 (Bishop et al., 1975, p. 517). Consequently the distribution of $\mathbf{W}^*(D_{\boldsymbol{\pi}^*})^{-1}\mathbf{W}^{*\prime}$ is the same as $\sum_{i=1}^{k-s-1} Z_i^2$, which is chi-squared with $k - s - 1$ degrees of freedom (since the Z_is are independent). From result (ii*) this is also the asymptotic distribution of $X^2 = \mathbf{W}_n^*(D_{\boldsymbol{\pi}^*})^{-1}\mathbf{W}_n^{*\prime} + o_p(1)$. □

Finally, we generalize result (4.4) in Section 4.1 to show:

Theorem A6.1. *Assuming* $\mathbf{X}$ *is a multinomial* $\text{Mult}_k(n, \boldsymbol{\pi}^*)$ *random vector, then*

$$2nI^\lambda(\mathbf{X}/n : \hat{\boldsymbol{\pi}}) = 2nI^1(\mathbf{X}/n : \hat{\boldsymbol{\pi}}) + o_p(1); \qquad -\infty < \lambda < \infty,$$

provided $\hat{\boldsymbol{\pi}} = \boldsymbol{\pi}^* + O_p(n^{-1/2})$.

PROOF.

$$\begin{aligned}2nI^\lambda(\mathbf{X}/n : \hat{\boldsymbol{\pi}}) &= \frac{2n}{\lambda(\lambda+1)}\sum_{i=1}^k \frac{X_i}{n}\left[\left(\frac{X_i}{n\hat{\pi}_i}\right)^\lambda - 1\right]\\ &= \frac{2n}{\lambda(\lambda+1)}\sum_{i=1}^k \hat{\pi}_i[(1 + V_i)^{\lambda+1} - 1] \qquad \text{(A6.4)}\end{aligned}$$

where $V_i = (X_i/n - \hat{\pi}_i)/\hat{\pi}_i$. By assumption, $\hat{\boldsymbol{\pi}} = \boldsymbol{\pi}^* + O_p(n^{-1/2})$ and from result (i) of Section A4 we know $\mathbf{X}/n = \boldsymbol{\pi}^* + O_p(n^{-1/2})$; therefore

$$V_i = O_p(n^{-1/2})/(\pi_i^* + O_p(n^{-1/2})) = O_p(n^{-1/2}),$$

provided $\pi_i^* > 0$. Expanding (A6.4) in a Taylor series gives

$$2nI^{\lambda}(\mathbf{X}/n : \hat{\boldsymbol{\pi}}) = \frac{2n}{\lambda(\lambda+1)} \sum_{i=1}^{k} \hat{\pi}_i \left[(\lambda+1)V_i + \frac{\lambda(\lambda+1)}{2} V_i^2 + O_p(n^{-3/2}) \right]$$
$$= n \sum_{i=1}^{k} \hat{\pi}_i V_i^2 + O_p(n^{-1/2})$$
$$= 2nI^{1}(\mathbf{X}/n : \hat{\boldsymbol{\pi}}) + O_p(n^{-1/2}),$$

as required. This ends the proof. □

When $\hat{\boldsymbol{\pi}}$ is BAN, the condition of Theorem A6.1 is immediately satisfied, and from result (iii*) it follows that $2nI^{\lambda}(\mathbf{X}/n : \hat{\boldsymbol{\pi}})$ has an asymptotic chi-squared distribution with $k - s - 1$ degrees of freedom.

A7. The Power-Divergence Generalization of the Chernoff-Lehmann Statistic: An Outline (Section 4.1)

Consider observing values of the random variables $Y_1, Y_2, \ldots, Y_n$ with the purpose of testing the hypothesis

$$H_0: \{Y_j\} \text{ have a common density function } g(y; \boldsymbol{\theta}), \tag{A7.1}$$

where $\boldsymbol{\theta} = (\theta_1, \theta_2, \ldots, \theta_s)$ must be estimated from the sample (Section 4.1). To perform a goodness-of-fit test of (A7.1), the Y_js must be grouped into k exhaustive and disjoint cells with frequencies $\mathbf{X} = (X_1, X_2, \ldots, X_k)$, $\sum_{i=1}^{k} X_i = n$; the vector $\mathbf{X}$ has the multinomial distribution $\text{Mult}_k(n, \boldsymbol{\pi})$. If $\{c_1, c_2, \ldots, c_k\}$ represents the disjoint partition of the domain of $g(y; \boldsymbol{\theta})$, which generates the k cells, then

$$\pi_i = f_i(\boldsymbol{\theta}) = \int_{c_i} g(y; \boldsymbol{\theta}) dy; \qquad i = 1, \ldots, k. \tag{A7.2}$$

To test (A7.1) using the power-divergence statistic $2nI^{\lambda}(\mathbf{X}/n : \boldsymbol{\pi})$ (defined in (4.1)), it is necessary to estimate $\boldsymbol{\theta}$, and hence $\boldsymbol{\pi}$ defined in (A7.2). Instead of estimating $\boldsymbol{\theta}$ from the grouped $\mathbf{X}$ (which leads to the standard results discussed in Sections A5 and A6), we shall consider estimating $\boldsymbol{\theta}$ from the ungrouped Y_js. For example, the maximum likelihood estimator (MLE) of $\boldsymbol{\theta}$ is obtained by choosing the value $\hat{\hat{\boldsymbol{\theta}}}$ that maximizes the likelihood

$$\prod_{j=1}^{n} g(Y_j; \boldsymbol{\theta}).$$

Generally speaking, the estimator $\hat{\hat{\boldsymbol{\theta}}}$ will not be BAN (i.e., will not satisfy (A5.4)) and therefore the power-divergence statistic $2nI^{\lambda}(\mathbf{X}/n : \hat{\hat{\boldsymbol{\pi}}})$, where $\hat{\hat{\boldsymbol{\pi}}} = \mathbf{f}(\hat{\hat{\boldsymbol{\theta}}})$, may not have the asymptotic chi-squared distribution with $k - s - 1$ degrees of freedom (Section A6). Chernoff and Lehmann (1954) derive the asymptotic distribution of

$$X^2 = \sum_{i=1}^{k} \frac{(X_i - n\hat{\hat{\pi}}_i)^2}{n\hat{\hat{\pi}}_i}, \tag{A7.3}$$

where $\hat{\hat{\boldsymbol{\pi}}}$ is the MLE based on the Y_js. The statistic in (A7.3) is called the Chernoff-Lehmann statistic. Assuming certain regularity conditions on $g(y; \boldsymbol{\theta})$ and $\mathbf{f}(\boldsymbol{\theta})$, Chernoff and Lehmann show that the asymptotic distribution of (A7.3) is equivalent to that of the random variable

$$\chi^2_{k-s-1} + \sum_{i=1}^{s} \zeta_i Z_i^2. \tag{A7.4}$$

Here χ^2_{k-s-1} is a random variable having a chi-squared distribution with $k - s - 1$ degrees of freedom, independent of the Z_i^2s, which are themselves independent chi-squared random variables, each with one degree of freedom. Each ζ_i is a constant in $[0, 1]$.

Using Theorem A6.1, this result can be generalized easily to the power-divergence statistic $2nI^\lambda(\mathbf{X}/n : \hat{\hat{\boldsymbol{\pi}}})$. Therefore provided $\hat{\hat{\boldsymbol{\pi}}} = \boldsymbol{\pi}^* + O_p(n^{-1/2})$, where $\boldsymbol{\pi}^*$ is the true value of $\boldsymbol{\pi}$ under H_0, the distributional result (A7.4) can be proved for general λ under the same conditions assumed by Chernoff and Lehmann (1954) in the case $\lambda = 1$, and the constants $\{\zeta_i\}$ do not depend on λ.

A8. Derivation of the Asymptotic Noncentral Chi-Squared Distribution for the Power-Divergence Statistic under Local Alternative Models (Section 4.2)

Consider the hypotheses (4.13)

$$H_{1,n}: \boldsymbol{\pi} = \boldsymbol{\pi}^* + \boldsymbol{\delta}/\sqrt{n}, \tag{A8.1}$$

where $\boldsymbol{\pi}^* \equiv \mathbf{f}(\boldsymbol{\theta}^*) \in \Pi_0$ is the true (but in general, unknown) value of $\boldsymbol{\pi}$ under H_0: $\boldsymbol{\pi} \in \Pi_0$ (Section A6) and $\sum_{i=1}^{k} \delta_i = 0$.

Using a similar argument and set of assumptions to those employed under H_0 in Section A6, we can show that (provided $\hat{\boldsymbol{\pi}} = \mathbf{f}(\hat{\boldsymbol{\theta}})$ is BAN under H_0: $\boldsymbol{\pi} \in \Pi_0$) the power-divergence statistic $2nI^\lambda(\mathbf{X}/n : \hat{\boldsymbol{\pi}})$ has an asymptotic noncentral chi-squared distribution under the hypotheses $H_{1,n}$. The degrees of freedom are the same as under H_0 (i.e., $k - s - 1$), and the noncentrality parameter is $\sum_{i=1}^{k} \delta_i^2/\pi_i^*$.

In the special case of Pearson's X^2 statistic ($\lambda = 1$), Mitra (1958) proves the required result (under Birch's regularity conditions). Due to the length of this proof, the interested reader is referred to Mitra's paper. We shall concentrate on showing how this result extends to other values of λ.

Theorem A8.1. *Assuming Birch's regularity conditions (Section A5) and assuming* $\hat{\boldsymbol{\theta}}$ *is a BAN estimator of* $\boldsymbol{\theta}^* \in \Theta_0$, *then*

$$2nI^\lambda(\mathbf{X}/n : \hat{\boldsymbol{\pi}}) = 2nI^1(\mathbf{X}/n : \hat{\boldsymbol{\pi}}) + o_p(1) \tag{A8.2}$$

as $n \to \infty$ under both H_0 and $H_{1,n}$ in (A8.1).

PROOF. The result is already proved under H_0 by Theorem A6.1, where it was shown that the conditions required for (A8.2) are

$$\begin{aligned} \mathbf{X}/n &= \boldsymbol{\pi}^* + O_p(n^{-1/2}) \\ \hat{\boldsymbol{\pi}} &= \boldsymbol{\pi}^* + O_p(n^{-1/2}). \end{aligned} \tag{A8.3}$$

We shall now prove (A8.3) under the hypotheses $H_{1,n}$, where $\mathbf{X}$ is the multinomial $\text{Mult}_k(n, \boldsymbol{\pi}^* + \boldsymbol{\delta}/\sqrt{n})$ random vector.

Paralleling result (i) of Section A4 we know that under $H_{1,n}$, $\sqrt{n}(\mathbf{X}/n - \boldsymbol{\pi}^* - \boldsymbol{\delta}/\sqrt{n})$ has an asymptotic multivariate normal distribution, and therefore $\mathbf{X}/n - \boldsymbol{\pi}^* - \boldsymbol{\delta}/\sqrt{n} = O_p(n^{-1/2})$, which gives

$$\mathbf{X}/n - \boldsymbol{\pi}^* = O_p(n^{-1/2}). \tag{A8.4}$$

Now recall (Section A5) that the BAN estimator $\hat{\boldsymbol{\theta}}$ of $\boldsymbol{\theta}^*$ satisfies

$$\hat{\boldsymbol{\theta}} = \boldsymbol{\theta}^* + (\mathbf{X}/n - \boldsymbol{\pi}^*)(D_{\boldsymbol{\pi}^*})^{-1/2}A(A'A)^{-1} + o_p(n^{-1/2}).$$

From (A8.4) it follows that $\hat{\boldsymbol{\theta}} - \boldsymbol{\theta}^* = O_p(n^{-1/2})$, and therefore Birch's regularity condition (c) (Section A5) gives

$$\begin{aligned} \mathbf{f}(\hat{\boldsymbol{\theta}}) - \mathbf{f}(\boldsymbol{\theta}^*) &= (\hat{\boldsymbol{\theta}} - \boldsymbol{\theta}^*)(\partial \mathbf{f}(\boldsymbol{\theta}^*)/\partial \boldsymbol{\theta})' + o(|\hat{\boldsymbol{\theta}} - \boldsymbol{\theta}^*|) \\ &= O_p(n^{-1/2}). \end{aligned}$$

Consequently $\hat{\boldsymbol{\pi}} - \boldsymbol{\pi}^* = O_p(n^{-1/2})$, which completes the proof. □

A9. Derivation of the Mean and Variance of the Power-Divergence Statistic for $\lambda > -1$ under a Nonlocal Alternative Model (Section 4.2)

Let $\mathbf{X}$ be a multinomial $\text{Mult}_k(n, \boldsymbol{\pi})$ random vector and consider the simple null and alternative hypotheses

$$H_0: \boldsymbol{\pi} = \boldsymbol{\pi}_0$$

versus

$$H_1: \boldsymbol{\pi} = \boldsymbol{\pi}_1,$$

where both $\boldsymbol{\pi}_0$ and $\boldsymbol{\pi}_1$ are completely specified and $\boldsymbol{\pi}_1$ does not depend on n. Assume further that all the elements of $\boldsymbol{\pi}_0$ and $\boldsymbol{\pi}_1$ are positive.

From (4.1) we can write $2nI^\lambda(\mathbf{X}/n : \boldsymbol{\pi}_0)/n$ as

$$
\begin{aligned}
2I^\lambda(\mathbf{X}/n:\boldsymbol{\pi}_0) &= \frac{2}{\lambda(\lambda+1)}\sum_{i=1}^{k}\pi_{0i}\left[\left(\frac{X_i}{n\pi_{0i}}\right)^{\lambda+1}-1\right]\\
&= \frac{2}{\lambda(\lambda+1)}\sum_{i=1}^{k}\pi_{0i}\left[\frac{(X_i/n)^{\lambda+1}-\pi_{1i}^{\lambda+1}}{\pi_{0i}^{\lambda+1}}+\frac{\pi_{1i}^{\lambda+1}-\pi_{0i}^{\lambda+1}}{\pi_{0i}^{\lambda+1}}\right]\\
&= \frac{2}{\lambda(\lambda+1)}\left\{\sum_{i=1}^{k}\pi_{1i}\left(\frac{\pi_{1i}}{\pi_{0i}}\right)^{\lambda}\left[\left(\frac{X_i}{n\pi_{1i}}\right)^{\lambda+1}-1\right]\right.\\
&\quad\left.+\sum_{i=1}^{k}\pi_{0i}\left[\left(\frac{\pi_{1i}}{\pi_{0i}}\right)^{\lambda+1}-1\right]\right\}\\
&= \frac{2}{\lambda(\lambda+1)}\sum_{i=1}^{k}\pi_{1i}\left(\frac{\pi_{1i}}{\pi_{0i}}\right)^{\lambda}\left[\left(1+\frac{W_{1i}}{\sqrt{n\pi_{1i}}}\right)^{\lambda+1}-1\right]+2I^\lambda(\boldsymbol{\pi}_1:\boldsymbol{\pi}_0),
\end{aligned}
$$

where $W_{1i}=\sqrt{n}(X_i/n-\pi_{1i})$. Assuming the alternative hypothesis H_1 is true, then W_{1i} is asymptotically normally distributed (Section A4), and the term involving W_{1i} in the preceding equation can be expanded in a Taylor series to give

$$
2I^\lambda(\mathbf{X}/n:\boldsymbol{\pi}_0)=\sum_{i=1}^{k}\left(\frac{\pi_{1i}}{\pi_{0i}}\right)^{\lambda}\left[\frac{2W_{1i}}{\lambda\sqrt{n}}+\frac{W_{1i}^2}{n\pi_{1i}}\right]+o_p(1/n)+2I^\lambda(\boldsymbol{\pi}_1:\boldsymbol{\pi}_0). \quad \text{(A9.1)}
$$

Noting that $E[W_{1i}]=0$ and $E[W_{1i}^2]=\pi_{1i}-\pi_{1i}^2$ in (A9.1), we see that

$$
E[2I^\lambda(\mathbf{X}/n:\boldsymbol{\pi}_0)]=2I^\lambda(\boldsymbol{\pi}_1:\boldsymbol{\pi}_0)+O(1/n),
$$

provided $\lambda>-1$ (Bishop et al., 1975, pp. 486–488). Now consider

$$
\begin{aligned}
\operatorname{var}[2I^\lambda(\mathbf{X}/n:\boldsymbol{\pi}_0)] &= \sum_{i=1}^{k}\left(\frac{\pi_{1i}}{\pi_{0i}}\right)^{2\lambda}\operatorname{var}\left[\frac{2W_{1i}}{\lambda\sqrt{n}}+\frac{W_{1i}^2}{n\pi_{1i}}\right]\\
&\quad+\sum_{i\neq j}^{k}\left(\frac{\pi_{1i}\pi_{1j}}{\pi_{0i}\pi_{0j}}\right)^{\lambda}\operatorname{cov}\left[\left(\frac{2W_{1i}}{\lambda\sqrt{n}}+\frac{W_{1i}^2}{n\pi_{1i}}\right),\left(\frac{2W_{1j}}{\lambda\sqrt{n}}+\frac{W_{1j}^2}{n\pi_{1j}}\right)\right]\\
&= \frac{4}{n\lambda^2}\left[\sum_{i=1}^{k}\left(\frac{\pi_{1i}}{\pi_{0i}}\right)^{2\lambda}\operatorname{var}[W_{1i}]\right.\\
&\quad\left.+\sum_{i\neq j}^{k}\left(\frac{\pi_{1i}\pi_{1j}}{\pi_{0i}\pi_{0j}}\right)^{\lambda}\operatorname{cov}[W_{1i},W_{1j}]\right]+o(1/n), \qquad \text{(A9.2)}
\end{aligned}
$$

provided $\lambda>-1$. Using $\operatorname{var}[W_{1i}]=E[W_{1i}^2]-E^2[W_{1i}]=\pi_{1i}-\pi_{1i}^2$, and $\operatorname{cov}[W_{1i},W_{1j}]=E[W_{1i}W_{1j}]-E[W_{1i}]E[W_{1j}]=-\pi_{1i}\pi_{1j}$, equation (A9.2) becomes ($\lambda>-1$),

$$
\operatorname{var}[2I^\lambda(\mathbf{X}/n:\boldsymbol{\pi}_0)]=\frac{4}{n\lambda^2}\left[\sum_{i=1}^{k}\left(\frac{\pi_{1i}}{\pi_{0i}}\right)^{2\lambda}\pi_{1i}-\left(\sum_{i=1}^{k}\left(\frac{\pi_{1i}}{\pi_{0i}}\right)^{\lambda}\pi_{1i}\right)^2\right]+o(1/n).
$$

In the case of Pearson's X^2 statistic ($\lambda=1$), these results agree with those of Broffitt and Randles (1977).

A10. Proof of the Asymptotic Normality of the Power-Divergence Statistic under Sparseness Assumptions (Section 4.3)

Assume $\mathbf{X}_k = (X_{1k}, X_{2k}, \ldots, X_{kk})$ is a multinomial $\text{Mult}_k(n_k, \boldsymbol{\pi}_k)$ random vector. To test the hypotheses

$$H_0\colon \boldsymbol{\pi}_k = \boldsymbol{\pi}_{0k},$$

consider the test statistic of the form

$$S_k = \sum_{i=1}^{k} h_k(X_{ik}, i/k), \tag{A10.1}$$

where $h_k(\cdot, \cdot)$ is a real measurable function. We quote the following result due to Holst (1972).

Theorem A10.1. *Define*

$$\begin{aligned} \mu_k &= \sum_{i=1}^{k} E[h_k(Y_{ik}, i/k)] \\ \sigma_k^2 &= \sum_{i=1}^{k} \text{var}[h_k(Y_{ik}, i/k)] - \left\{\sum_{i=1}^{k} \text{cov}[Y_{ik}, h_k(Y_{ik}, i/k)]\right\}^2 \Big/ n, \end{aligned} \tag{A10.2}$$

where the Y_{ik}s are independent Poisson random variables with means $n_k \pi_{0ik}$; $i = 1, \ldots, k$. Assume

(a) n_k *and* $k \to \infty$ *such that* $n_k/k \to a$ $(0 < a < \infty)$;
(b) $k\pi_{0ik} \le c < \infty$ *for some nonnegative number* c; $i = 1, \ldots, k$ *and all* k;
(c) $|h_k(v, x)| \le \alpha \exp(\beta v)$ *for* $0 \le x \le 1$; $v = 0, 1, 2, \ldots$; α *and* β *real; and*
(d) $0 < \liminf_{n_k \to \infty} \sigma_k^2/n_k \le \limsup_{n_k \to \infty} \sigma_k^2/n_k < \infty$.

Then $(S_k - \mu_k)/\sigma_k$ *is asymptotically a standard normal random variable, as* $k \to \infty$.

This theorem indicates that while S_k in (A10.1) is a sum of dependent random variables (and hence the standard central limit theorems cannot be applied), under certain conditions it is possible to ensure that S_k has the same asymptotic limit as $S_k^* = \sum_{i=1}^{k} h_k(Y_{ik}, i/k)$. The Y_{ik}s are independent Poisson random variables with the same means as the multinomial X_{ik}s.

Cressie and Read (1984) apply this theorem in two interesting cases (further details are given by Read, 1982). First, consider the equiprobable hypothesis

$$H_0\colon \boldsymbol{\pi}_k = \mathbf{1}/k$$

(where $\mathbf{1} = (1, 1, \ldots, 1)$ is a vector of length k). Condition (b) of Theorem A10.1 is immediately satisfied, since $\pi_{0ik} = 1/k$ for each $i = 1, \ldots, k$. Now define

$$h_k(X_{ik}, i/k) = \begin{cases} \dfrac{2}{\lambda(\lambda + 1)} \dfrac{n_k}{k} \left[\left(\dfrac{X_{ik}}{n_k/k}\right)^{\lambda+1} - 1\right]; & \lambda \ne 0, \lambda > -1 \\ 2X_{ik} \log(X_{ik}k/n_k); & \lambda = 0, \end{cases} \tag{A10.3}$$

for any given $\lambda > -1$. Condition (c) of the theorem is clearly satisfied, and condition (d) can be shown to hold by appealing to lemma 2.3 of Holst (1972). Finally, we obtain the asymptotic normality of the power-divergence statistic (result (4.16)) by substituting expression (A10.3) for $h_k(\cdot, \cdot)$ into (A10.2), and noting that for the equiprobable hypothesis, the Y_{ik}s of Theorem A10.1 are independent and identically distributed Poisson random variables with mean n_k/k.

The second application of Theorem A10.1 is to the local alternative hypotheses (4.17)

$$H_{1,k}\colon \boldsymbol{\pi}_k = \mathbf{1}/k + \boldsymbol{\delta}/n_k^{1/4},$$

where $\delta_i = \int_{(i-1)/k}^{i/k} c(x)dx$, $\int_0^1 c(x)dx = 0$, and c is a known continuous function on $[0, 1]$. The details of applying Theorem A10.1 are similar to the null case (Read, 1982). From the resulting asymptotic normality of $2n_k I^\lambda(\mathbf{X}_k/n : \mathbf{1}/k)$ under both H_0 and $H_{1,k}$, we can derive the Pitman asymptotic efficiency of the test by evaluating the limit of the ratio $(\mu_{k,1}^{(\lambda)} - \mu_{k,0}^{(\lambda)})/\sigma_{k,1}^{(\lambda)}$. Here $\mu_{k,0}^{(\lambda)}$ represents the mean of $2n_k I^\lambda(\mathbf{X}_k/n : \mathbf{1}/k)$ under the equiprobable null hypothesis and $\mu_{k,1}^{(\lambda)}$ and $\sigma_{k,1}^{(\lambda)}$ represent the mean and standard deviation under the local alternatives described herein.

Paralleling the results of Holst (1972) and Ivchenko and Medvedev (1978), we obtain

$$(\mu_{k,1}^{(\lambda)} - \mu_{k,0}^{(\lambda)})/\sigma_{k,1}^{(\lambda)} \to \sqrt{(a/2)}\,\mathrm{sgn}(\lambda) \int_0^1 [c(x)]^2 dx \,\mathrm{corr}\{Y^{\lambda+1} - a^{-1}\mathrm{cov}(Y^{\lambda+1}, Y)Y,\ Y^2 - (2a+1)Y\}, \quad \text{(A10.4)}$$

where Y has a Poisson distribution with mean a. By noting that the correlation term in (A10.4) is 1 when $\lambda = 1$, it follows that Pearson's X^2 test ($\lambda = 1$) will obtain maximal efficiency among the power-divergence family for testing (4.17). Further analysis of (A10.4), for specific values of λ and a, is provided by Read (1982).

A11. Derivation of the First Three Moments (to Order $1/n$) of the Power-Divergence Statistic for $\lambda > -1$ under the Classical (Fixed-Cells) Assumptions (Section 5.1)

Consider the simple null hypothesis

$$H_0\colon \boldsymbol{\pi} = \boldsymbol{\pi}_0. \quad \text{(A11.1)}$$

From (4.5), it follows that the asymptotic distribution of the power-divergence statistic $2nI^\lambda(\mathbf{X}/n : \boldsymbol{\pi}_0)$ (defined from (4.1)) is chi-squared with $k - 1$ degrees of freedom. Consequently, the first three moments of $2nI^\lambda(\mathbf{X}/n : \boldsymbol{\pi}_0)$ are asymptotically equal to the first three moments of a chi-squared random variable with $k - 1$ degrees of freedom, provided $\lambda > -1$ (for $\lambda \le -1$ the moments do not exist; see Bishop et al., 1975, pp. 486–488).

$$E[2nI^{\lambda}(\mathbf{X}/n:\boldsymbol{\pi}_0)] \to k-1,$$
$$E[(2nI^{\lambda}(\mathbf{X}/n:\boldsymbol{\pi}_0))^2] \to k^2-1,$$
$$E[(2nI^{\lambda}(\mathbf{X}/n:\boldsymbol{\pi}_0))^3] \to k^3+3k^2-k-3,$$

as $n \to \infty$, with k and $\boldsymbol{\pi}_0$ fixed.

We now derive higher-order correction terms for the first three moments of $2nI^{\lambda}(\mathbf{X}/n:\boldsymbol{\pi}_0)$ as $n \to \infty$ (from Read, 1982), which are stated without proof in Cressie and Read (1984). Rewrite (4.1) as

$$2nI^{\lambda}(\mathbf{X}/n:\boldsymbol{\pi}_0) = \frac{2}{\lambda(\lambda+1)}\sum_{i=1}^{k} X_i\left[\left(\frac{X_i}{n\pi_{0i}}\right)^{\lambda}-1\right]$$
$$= \frac{2n}{\lambda(\lambda+1)}\sum_{i=1}^{k} \pi_{0i}\left[\left(1+\frac{W_i}{\sqrt{n\pi_{0i}}}\right)^{\lambda+1}-1\right], \quad \text{(A11.2)}$$

where $W_i = \sqrt{n}(X_i/n - \pi_{0i})$. Expanding (A11.2) in a Taylor series gives

$$2nI^{\lambda}(\mathbf{X}/n:\boldsymbol{\pi}_0) = \frac{2n}{\lambda(\lambda+1)}\sum_{i=1}^{k} \pi_{0i}\left[1+\frac{(\lambda+1)W_i}{\sqrt{n\pi_{0i}}}\right.$$
$$+\frac{\lambda(\lambda+1)W_i^2}{2(\sqrt{n\pi_{0i}})^2}+\frac{(\lambda-1)\lambda(\lambda+1)W_i^3}{6(\sqrt{n\pi_{0i}})^3}$$
$$\left.+\frac{(\lambda-2)(\lambda-1)\lambda(\lambda+1)W_i^4}{24(\sqrt{n\pi_{0i}})^4}+O_p(n^{-5/2})-1\right]$$
$$= \sum_{i=1}^{k}\frac{W_i^2}{\pi_{0i}}+\frac{\lambda-1}{3\sqrt{n}}\sum_{i=1}^{k}\frac{W_i^3}{\pi_{0i}^2}+\frac{(\lambda-2)(\lambda-1)}{12n}\sum_{i=1}^{k}\frac{W_i^4}{\pi_{0i}^3}+O_p(n^{-3/2}),$$
$$\text{(A11.3)}$$

since $\sum_{i=1}^{k} W_i = 0$. The fact that $(W_i/\sqrt{n})^j = O_p(n^{-5/2})$ for $j = 5, 6, 7, \ldots$ follows from the asymptotic normality of W_i under H_0 (Section A4). Henceforth assume $\lambda > -1$.

First Moment

Taking expectations in (A11.3) gives

$$E[2nI^{\lambda}(\mathbf{X}/n:\boldsymbol{\pi}_0)] = \sum_{i=1}^{k}\frac{E[W_i^2]}{\pi_{0i}}+\frac{\lambda-1}{3\sqrt{n}}\sum_{i=1}^{k}\frac{E[W_i^3]}{\pi_{0i}^2}$$
$$+\frac{(\lambda-2)(\lambda-1)}{12n}\sum_{i=1}^{k}\frac{E[W_i^4]}{\pi_{0i}^3}+O(n^{-3/2}), \quad \text{(A11.4)}$$

since $E[O_p(n^{-3/2})] = O(n^{-3/2})$.

To obtain the moments of W_i, we shall use moment-generating functions. Recall that the moment-generating function of a multinomial $\text{Mult}_k(n, \boldsymbol{\pi}_0)$ random vector $\mathbf{X}$ is

$$M_{\mathbf{X}}(\mathbf{v}) = E[\exp(\mathbf{v}\mathbf{X}')] = \left[\sum_{i=1}^{k} \pi_{0i}\exp(v_i)\right]^n,$$

and therefore the moment-generating function of $\mathbf{W} = \sqrt{n}(\mathbf{X}/n - \boldsymbol{\pi}_0)$ is

$$\begin{aligned} M_{\mathbf{W}}(\mathbf{v}) &= E[\exp(\mathbf{v}\mathbf{W}')] \\ &= \exp(-n^{1/2}\mathbf{v}\boldsymbol{\pi}_0')E[\exp(n^{-1/2}\mathbf{v}\mathbf{X}')] \\ &= \exp(-n^{1/2}\mathbf{v}\boldsymbol{\pi}_0')M_{\mathbf{X}}(n^{-1/2}\mathbf{v}). \end{aligned} \tag{A11.5}$$

Using (A11.5) we can obtain the moments of W_i from

$$E[W_i^a] = \frac{\partial^a}{\partial v_i^a} M_{\mathbf{W}}(\mathbf{v})\bigg|_{\mathbf{v}=\mathbf{0}}, \tag{A11.6}$$

for $i = 1, \ldots, k$; $a = 1, 2, 3, \ldots$.

Applying (A11.5) and (A11.6) gives

$$E[W_i^2] = -\pi_{0i}^2 + \pi_{0i},$$

$$E[W_i^3] = n^{-1/2}(2\pi_{0i}^3 - 3\pi_{0i}^2 + \pi_{0i}),$$

$$E[W_i^4] = 3\pi_{0i}^4 - 6\pi_{0i}^3 + 3\pi_{0i}^2 + n^{-1}(-6\pi_{0i}^4 + 12\pi_{0i}^3 - 7\pi_{0i}^2 + \pi_{0i}).$$

Substituting these formulas into (A11.4) gives the first moment to order n^{-1},

$$\begin{aligned} E[2nI^\lambda(\mathbf{X}/n : \boldsymbol{\pi}_0)] = k - 1 + \frac{1}{n}\bigg[&\frac{\lambda - 1}{3}(2 - 3k + t) \\ &+ \frac{(\lambda - 1)(\lambda - 2)}{4}(1 - 2k + t)\bigg] + O(n^{-3/2}), \end{aligned}$$

where $t = \sum_{i=1}^{k} \pi_{0i}^{-1}$.

Second Moment

Squaring (A11.3) and taking expectations gives

$$\begin{aligned} E[(2nI^\lambda(\mathbf{X}/n : \boldsymbol{\pi}_0))^2] = &\sum_{i=1}^{k} \frac{E[W_i^4]}{\pi_{0i}^2} + \sum_{i \neq j}^{k} \frac{E[W_i^2 W_j^2]}{\pi_{0i}\pi_{0j}} \\ &+ \frac{2(\lambda - 1)}{3\sqrt{n}}\left[\sum_{i=1}^{k} \frac{E[W_i^5]}{\pi_{0i}^3} + \sum_{i \neq j}^{k} \frac{E[W_i^2 W_j^3]}{\pi_{0i}\pi_{0j}^2}\right] \\ &+ \frac{1}{n}\left[\frac{(\lambda - 1)(\lambda - 2)}{6}\left(\sum_{i=1}^{k} \frac{E[W_i^6]}{\pi_{0i}^4} + \sum_{i \neq j}^{k} \frac{E[W_i^2 W_j^4]}{\pi_{0i}\pi_{0j}^3}\right)\right. \\ &\left.+ \frac{(\lambda - 1)^2}{9}\left(\sum_{i=1}^{k} \frac{E[W_i^6]}{\pi_{0i}^4} + \sum_{i \neq j}^{k} \frac{E[W_i^3 W_j^3]}{\pi_{0i}^2\pi_{0j}^2}\right)\right] + O(n^{-3/2}). \end{aligned} \tag{A11.7}$$

The joint moments of W_i and W_j can be obtained from

$$E[W_i^a W_j^b] = \frac{\partial^a}{\partial v_i^a} \frac{\partial^b}{\partial v_j^b} M_{\mathbf{W}}(\mathbf{v}) \bigg|_{\mathbf{v}=\mathbf{0}}, \tag{A11.8}$$

for $i, j = 1, \ldots, k$; $a, b = 1, 2, 3, \ldots$. Therefore, substituting (A11.5) into (A11.6) and (A11.8) gives

$$\begin{aligned} E[W_i^5] &= n^{-1/2}(-20\pi_{0i}^5 + 50\pi_{0i}^4 - 40\pi_{0i}^3 + 10\pi_{0i}^2) \\ &\quad + n^{-3/2}(24\pi_{0i}^5 - 60\pi_{0i}^4 + 50\pi_{0i}^3 - 15\pi_{0i}^2 + \pi_{0i}), \end{aligned}$$

and

$$\begin{aligned} E[W_i^6] &= -15\pi_{0i}^6 + 45\pi_{0i}^5 - 45\pi_{0i}^4 + 15\pi_{0i}^3 \\ &\quad + n^{-1}(130\pi_{0i}^6 - 390\pi_{0i}^5 + 415\pi_{0i}^4 - 180\pi_{0i}^3 + 25\pi_{0i}^2) + O(n^{-2}). \end{aligned}$$

For $i \neq j$ both fixed, define $\pi_{ab} = \pi_{0i}^a \pi_{0j}^b$, then

$$\begin{aligned} E[W_i^2 W_j^2] &= 3\pi_{22} - \pi_{21} - \pi_{12} + \pi_{11} \\ &\quad + n^{-1}(-6\pi_{22} + 2\pi_{21} + 2\pi_{12} - \pi_{11}), \\ E[W_i^2 W_j^3] &= n^{-1/2}(-20\pi_{23} + 5\pi_{13} + 15\pi_{22} - 6\pi_{12} - \pi_{21} + \pi_{11}), \\ E[W_i^2 W_j^4] &= -15\pi_{24} + 18\pi_{23} + 3\pi_{14} - 3\pi_{22} - 6\pi_{13} + 3\pi_{12} \\ &\quad + n^{-1}(130\pi_{24} - 156\pi_{23} - 26\pi_{14} + 41\pi_{22} + 42\pi_{13} \\ &\quad - \pi_{21} - 17\pi_{12} + \pi_{11}) + O(n^{-2}), \end{aligned}$$

and

$$\begin{aligned} E[W_i^3 W_j^3] &= -15\pi_{33} + 9\pi_{32} + 9\pi_{23} - 9\pi_{22} \\ &\quad + n^{-1}(130\pi_{33} - 78\pi_{32} - 78\pi_{23} + 63\pi_{22} + 5\pi_{31} + 5\pi_{13} \\ &\quad - 6\pi_{21} - 6\pi_{12} + \pi_{11}) + O(n^{-2}). \end{aligned}$$

Substituting these formulas into (A11.7) and simplifying gives the second moment to order n^{-1}, viz.,

$$\begin{aligned} E[(2nI^\lambda(\mathbf{X}/n : \boldsymbol{\pi}_0))^2] &= k^2 - 1 + \frac{1}{n}\bigg[(2 - 2k - k^2 + t) \\ &\quad + \frac{2(\lambda - 1)}{3}(10 - 13k - 6k^2 + (k + 8)t) \\ &\quad + \frac{(\lambda - 1)^2}{3}(4 - 6k - 3k^2 + 5t) \\ &\quad + \frac{(\lambda - 1)(\lambda - 2)}{2}(3 - 5k - 2k^2 + (k + 3)t)\bigg] \\ &\quad + O(n^{-3/2}); \end{aligned}$$

recall $t = \sum_{i=1}^{k} \pi_{0i}^{-1}$. This simplification uses the identities

$$\sum_{i=1}^{k} \pi_{0i}^2 + \sum_{i \neq j}^{k} \pi_{0i}\pi_{0j} = 1; \qquad \sum_{i \neq j}^{k} \pi_{0i} = k - 1; \qquad \text{and} \quad \sum_{i \neq j}^{k} \frac{\pi_{0i}}{\pi_{0j}} = \sum_{i=1}^{k} \frac{1}{\pi_{0i}} - k.$$

Third Moment

Cubing (A11.3) and taking expectations gives

$$\begin{aligned}
E[(2nI^{\lambda}(\mathbf{X}/n : \boldsymbol{\pi}_0))^3] &= \sum_{i=1}^{k} \frac{E[W_i^6]}{\pi_{0i}^3} + 3\sum_{i \neq j}^{k} \frac{E[W_i^4 W_j^2]}{\pi_{0i}^2 \pi_{0j}} + \sum_{\substack{i \neq j \neq l \\ i \neq l}}^{k} \frac{E[W_i^2 W_j^2 W_l^2]}{\pi_{0i}\pi_{0j}\pi_{0l}} \\
&\quad + \frac{\lambda - 1}{\sqrt{n}} \left[\sum_{i=1}^{k} \frac{E[W_i^7]}{\pi_{0i}^4} + \sum_{i \neq j}^{k} \frac{E[W_i^4 W_j^3]}{\pi_{0i}^2 \pi_{0j}^2} \right. \\
&\quad \left. + 2\sum_{i \neq j}^{k} \frac{E[W_i^2 W_j^5]}{\pi_{0i}\pi_{0j}^3} + \sum_{\substack{i \neq j \neq l \\ i \neq l}}^{k} \frac{E[W_i^2 W_j^2 W_l^3]}{\pi_{0i}\pi_{0j}\pi_{0l}^2} \right] \\
&\quad + \frac{1}{n} \left[\frac{(\lambda - 1)(\lambda - 2)}{4} \left(\sum_{i=1}^{k} \frac{E[W_i^8]}{\pi_{0i}^5} + \sum_{i \neq j}^{k} \frac{E[W_i^4 W_j^4]}{\pi_{0i}^2 \pi_{0j}^3} \right. \right. \\
&\quad \left. + 2\sum_{i \neq j}^{k} \frac{E[W_i^2 W_j^6]}{\pi_{0i}\pi_{0j}^4} + \sum_{\substack{i \neq j \neq l \\ i \neq l}}^{k} \frac{E[W_i^2 W_j^2 W_l^4]}{\pi_{0i}\pi_{0j}\pi_{0l}^3} \right) \\
&\quad + \frac{(\lambda - 1)^2}{3} \left(\sum_{i=1}^{k} \frac{E[W_i^8]}{\pi_{0i}^5} + \sum_{i \neq j}^{k} \frac{E[W_i^6 W_j^2]}{\pi_{0i}^4 \pi_{0j}} \right. \\
&\quad \left. \left. + 2\sum_{i \neq j}^{k} \frac{E[W_i^3 W_j^5]}{\pi_{0i}^2 \pi_{0j}^3} + \sum_{\substack{i \neq j \neq l \\ i \neq l}}^{k} \frac{E[W_i^3 W_j^3 W_l^2]}{\pi_{0i}^2 \pi_{0j}^2 \pi_{0l}} \right) \right] \\
&\quad + O(n^{-3/2}).
\end{aligned} \tag{A11.9}$$

Therefore substituting (A11.5) into (A11.6) and (A11.8) gives

$$E[W_i^7] = n^{-1/2}(210\pi_{0i}^7 - 735\pi_{0i}^6 + 945\pi_{0i}^5 - 525\pi_{0i}^4 + 105\pi_{0i}^3) + O(n^{-3/2}),$$

and

$$E[W_i^8] = 105\pi_{0i}^8 - 420\pi_{0i}^7 + 630\pi_{0i}^6 - 420\pi_{0i}^5 + 105\pi_{0i}^4 + O(n^{-1}).$$

For $i \neq j \neq l$ and $i \neq l$ all fixed, define $\pi_{ab} = \pi_{0i}^a \pi_{0j}^b$ and $\pi_{abc} = \pi_{0i}^a \pi_{0j}^b \pi_{0l}^c$; then

$$\begin{aligned}
E[W_i^4 W_j^3] &= n^{-1/2}(210\pi_{43} - 105\pi_{42} - 210\pi_{33} + 3\pi_{41} + 144\pi_{32} \\
&\quad + 36\pi_{23} - 6\pi_{31} - 39\pi_{22} + 3\pi_{21}) + O(n^{-3/2}), \\
E[W_i^4 W_j^4] &= 105\pi_{44} - 90(\pi_{43} + \pi_{34}) + 9(\pi_{24} + \pi_{42}) + 108\pi_{33} \\
&\quad - 18(\pi_{32} + \pi_{23}) + 9\pi_{22} + O(n^{-1}),
\end{aligned}$$

$$E[W_i^5 W_j^2] = n^{-1/2}(210\pi_{52} - 35\pi_{51} - 350\pi_{42} + 80\pi_{41} + 150\pi_{32} - 55\pi_{31} - 10\pi_{22} + 10\pi_{21}) + O(n^{-3/2}),$$

$$E[W_i^5 W_j^3] = 105\pi_{53} - 45\pi_{52} - 150\pi_{43} + 90\pi_{42} + 45\pi_{33} - 45\pi_{32} + O(n^{-1}),$$

$$E[W_i^6 W_j^2] = 105\pi_{62} - 15\pi_{61} - 225\pi_{52} + 45\pi_{51} + 135\pi_{42} - 45\pi_{41} - 15\pi_{32} + 15\pi_{31} + O(n^{-1}),$$

$$E[W_i^2 W_j^2 W_l^2] = -15\pi_{222} + 3(\pi_{122} + \pi_{212} + \pi_{221}) - (\pi_{112} + \pi_{121} + \pi_{211}) + \pi_{111} + n^{-1}[130\pi_{222} - 26(\pi_{122} + \pi_{212} + \pi_{221}) + 7(\pi_{112} + \pi_{121} + \pi_{211}) - 3\pi_{111}] + O(n^{-2}),$$

$$E[W_i^2 W_j^2 W_l^3] = n^{-1/2}[210\pi_{223} - 105\pi_{222} - 35(\pi_{123} + \pi_{213}) + 8\pi_{113} + 24(\pi_{122} + \pi_{212}) + 3\pi_{221} - 9\pi_{112} - (\pi_{211} + \pi_{121}) + \pi_{111}] + O(n^{-3/2}),$$

$$E[W_i^2 W_j^3 W_l^3] = 105\pi_{233} - 15\pi_{133} - 45(\pi_{223} + \pi_{232}) + 9(\pi_{132} + \pi_{123}) + 27\pi_{222} - 9\pi_{122} + O(n^{-1}),$$

and

$$E[W_i^2 W_j^2 W_l^4] = 105\pi_{224} - 15(\pi_{124} + \pi_{214}) - 90\pi_{223} + 3\pi_{114} + 18(\pi_{123} + \pi_{213}) + 9\pi_{222} - 6\pi_{113} - 3(\pi_{122} + \pi_{212}) + 3\pi_{112} + O(n^{-1}).$$

Substituting all these formulas into (A11.9) gives the third moment to order n^{-1},

$$\begin{aligned} E[2nI^{\lambda}(\mathbf{X}/n : \boldsymbol{\pi}_0))^3] = {} & k^3 + 3k^2 - k - 3 \\ & + n^{-1}[26 - 24k - 21k^2 - 3k^3 + (19 + 3k)t \\ & + (\lambda - 1)(70 - 81k - 64k^2 - 9k^3 + (65 + 18k + k^2)t) \\ & + (\lambda - 1)^2(20 - 26k - 21k^2 - 3k^3 + (25 + 5k)t) \\ & + \frac{3(\lambda - 1)(\lambda - 2)}{4}(15 - 22k - 15k^2 - 2k^3 \\ & + (15 + 8k + k^2)t)] + O(n^{3/2}); \end{aligned}$$

recall $t = \sum_{i=1}^{k} \pi_{0i}^{-1}$. This simplification uses the identities

$$\sum_{i \neq j}^{k} \frac{\pi_{0i}^2}{\pi_{0j}} + \sum_{\substack{i \neq j \neq l \\ i \neq l}}^{k} \frac{\pi_{0i}\pi_{0j}}{\pi_{0l}} = \sum_{i=1}^{k} \frac{1}{\pi_{0i}} - 2k + 1,$$

and

$$\sum_{i=1}^{k} \pi_{0i}^3 + 3 \sum_{i \neq j}^{k} \pi_{0i}^2 \pi_{0j} + \sum_{\substack{i \neq j \neq l \\ i \neq l}}^{k} \pi_{0i} \pi_{0j} \pi_{0l} = 1.$$

The expressions for the first two moments agree with those of Johnson and Kotz (1969, p. 286) for Pearson's X^2 ($\lambda = 1$) (originally derived by Haldane, 1937) and Smith et al. (1981) for the loglikelihood ratio statistic G^2 ($\lambda = 0$).

For all three moments, the first-order terms are the moments of a chi-squared random variable with $k - 1$ degrees of freedom. The second-order (i.e., order n^{-1}) terms can be considered correction terms for any given $\lambda > -1$. Cressie and Read (1984) calculate which values of λ need no correction for specified values of k and t. In the cases they consider, they find the solutions for λ to be in the range [0.32, 2.68]. One of their more interesting conclusions is that for larger k (say $k > 20$), the solutions for λ settle at $\lambda = 1$ and $\lambda = 2/3$ (Cressie and Read, 1984, table 3).

A12. Derivation of the Second-Order Terms for the Distribution Function of the Power-Divergence Statistic under the Classical (Fixed-Cells) Assumptions (Section 5.2)

In this section we use the classical (fixed-cells) assumptions, which lead to the asymptotic chi-squared limit for the distribution function of the power-divergence statistic. We derive the expressions for the second-order terms (given by (5.9) and (5.10)) in the expansion of the distribution function. More details of these results can be found in Yarnold (1972), Siotani and Fujikoshi (1984), and Read (1984a).

Let $\mathbf{X}$ be a multinomial $\text{Mult}_k(n, \boldsymbol{\pi}_0)$ random row vector (which has dimension k). Define $W_i = \sqrt{n}(X_i/n - \pi_{0i})$; $i = 1, \ldots, k$ and let $\tilde{\mathbf{W}} = (W_1, W_2, \ldots, W_r)$ where $r = k - 1$. Henceforth we use the convention that "$\tilde{}$" above the vector refers to the $(k-1)$-dimensional version. Therefore $\tilde{\mathbf{W}}$ is a random row vector of dimension $k - 1$, taking values in the lattice

$$L = \{\tilde{\mathbf{w}} = (w_1, w_2, \ldots, w_r): \tilde{\mathbf{w}} = \sqrt{n}(\tilde{\mathbf{x}}/n - \tilde{\boldsymbol{\pi}}_0) \text{ and } \tilde{\mathbf{x}} \in M\}, \quad \text{(A12.1)}$$

where $\tilde{\boldsymbol{\pi}}_0 = (\pi_{01}, \pi_{02}, \ldots, \pi_{0r})$ and $M = \{\tilde{\mathbf{x}} = (x_1, x_2, \ldots, x_r)$: x_i is a nonnegative integer; $i = 1, \ldots, r$; $\sum_{i=1}^r x_i \leq n\}$. Then the asymptotic expansion of the probability mass function for $\tilde{\mathbf{W}}$ is given, for any $\tilde{\mathbf{w}} \in L$, by

$$Pr(\tilde{\mathbf{W}} = \tilde{\mathbf{w}}) = n^{-r/2}\phi(\tilde{\mathbf{w}})[1 + n^{-1/2}h_1(\tilde{\mathbf{w}}) + n^{-1}h_2(\tilde{\mathbf{w}}) + O(n^{-3/2})] \quad \text{(A12.2)}$$

(Siotani and Fujikoshi, 1984), where $\phi(\tilde{\mathbf{w}})$ is the multivariate normal density function

$$\phi(\tilde{\mathbf{w}}) = (2\pi)^{-r/2}|\Omega|^{-1/2}\exp(-\tilde{\mathbf{w}}\Omega^{-1}\tilde{\mathbf{w}}'/2),$$

and

$$h_1(\tilde{\mathbf{w}}) = -\frac{1}{2}\sum_{i=1}^{k}\frac{w_i}{\pi_{0i}} + \frac{1}{6}\sum_{i=1}^{k}\frac{w_i^3}{\pi_{0i}^2},$$

$$h_2(\tilde{\mathbf{w}}) = \frac{1}{2}[h_1(\tilde{\mathbf{w}})]^2 + \frac{1}{12}\left(1 - \sum_{i=1}^{k}\frac{1}{\pi_{0i}}\right) + \frac{1}{4}\sum_{i=1}^{k}\frac{w_i^2}{\pi_{0i}^2} - \frac{1}{12}\sum_{i=1}^{k}\frac{w_i^4}{\pi_{0i}^3},$$

with $w_k = -\sum_{i=1}^{r} w_i$, and $\Omega = \text{diag}(\pi_{01}, \pi_{02}, \ldots, \pi_{0r}) - \tilde{\boldsymbol{\pi}}_0'\tilde{\boldsymbol{\pi}}_0$.

The expansion (A12.2) is a special case of Yarnold's (1972) equation (1.6) evaluated for the random vector $\tilde{\mathbf{W}}$, and provides a local Edgeworth approximation for the probability of $\tilde{\mathbf{W}}$ at each point $\tilde{\mathbf{w}}$ in the lattice (A12.1). If $\tilde{\mathbf{W}}$ had a continuous distribution function, we could use the standard Edgeworth expansion,

$$Pr(\tilde{\mathbf{W}} \in B) = \int \cdots \int_B \phi(\tilde{\mathbf{w}})[1 + n^{-1/2}h_1(\tilde{\mathbf{w}}) + n^{-1}h_2(\tilde{\mathbf{w}})]d\tilde{\mathbf{w}} + O(n^{-3/2}), \quad \text{(A12.3)}$$

to calculate the probability of any set B. However $\tilde{\mathbf{W}}$ has a lattice distribution and Yarnold (1972) indicates that in this case, the expansion (A12.3) is not valid. To obtain a valid expansion for $Pr(\tilde{\mathbf{W}} \in B)$, it is necessary to sum the local expansion (A12.2) over the set B. When B is an extended convex set (i.e., a convex set whose sections parallel to each coordinate axis are all intervals), Yarnold shows that we can write

$$Pr(\tilde{\mathbf{W}} \in B) = J_1 + J_2 + J_3 + O(n^{-3/2}), \quad \text{(A12.4)}$$

where J_1 is simply the Edgeworth expansion in (A12.3), J_2 is a discontinuous term to account for the discreteness of $\tilde{\mathbf{W}}$ (and is $O(n^{-1/2})$), and the J_3 term is a complicated term of $O(n^{-1})$ (Read, 1984a, section 2).

We can apply (A12.4) to the power-divergence family by observing that

$$Pr[2nI^{\lambda}(\mathbf{X}/n : \boldsymbol{\pi}_0) < c] = Pr[\tilde{\mathbf{W}} \in B_{\lambda}(c)],$$

where $B_{\lambda}(c) = \{\tilde{\mathbf{w}}: \tilde{\mathbf{w}} \in L \text{ and } 2nI^{\lambda}[(\tilde{\mathbf{w}}/\sqrt{n} + \tilde{\boldsymbol{\pi}}_0, w_k/\sqrt{n} + \pi_{0k}): \boldsymbol{\pi}_0] < c\}$, for $w_k = -\sum_{i=1}^{r} w_i$.

Read (1984a) shows that $B_{\lambda}(c)$ is an extended convex set, and evaluates the terms J_1 and J_2 from (A12.4) when $B = B_{\lambda}(c)$ to give (5.9) and (5.10). The evaluation of J_1 relies on interpreting this term as a distribution function and simplifying the corresponding characteristic function to obtain a sum of four chi-squared characteristic functions with $k - 1$, $k + 1$, $k + 3$, and $k + 5$ degrees of freedom; the coefficients of these chi-squared characteristic functions are given by (5.9). J_2 is evaluated to $o(n^{-1})$ using the results of Yarnold (1972, p. 1572). The term J_3 is very complicated, but Read points out that any λ-dependent terms will be $O(n^{-3/2})$, and hence from the point of view of (A12.4), J_3 does not depend on λ. In the special cases of Pearson's X^2 ($\lambda = 1$), the loglikelihood ratio statistic G^2 ($\lambda = 0$), and the Freeman-Tukey statistic F^2 ($\lambda = -1/2$), equations (5.9) and (5.10) coincide with the results of Yarnold (1972) and Siotani and Fujikoshi (1984).

A13. Derivation of the Minimum Asymptotic Value of the Power-Divergence Statistic (Section 6.2)

Given a single fixed observed cell frequency $x_i = \delta$ for some fixed $j\,(1 \leq j \leq k)$, the minimum asymptotic value of $2I^\lambda(\mathbf{x} : \mathbf{m})$ (from (2.17)) is defined as the smallest value of $2I^\lambda(\mathbf{x} : \mathbf{m})$ (under the constraint $x_j = \delta$) as $n \to \infty$.

The minimum asymptotic value of $2I^\lambda(\mathbf{x} : \mathbf{m})$ for $x_j = \delta\ (0 \leq \delta < n)$ is given by (6.4) to be

$$\frac{2}{\lambda(\lambda+1)}\left\{\delta\left[\left(\frac{\delta}{m_j}\right)^\lambda - 1\right] + \lambda(m_j - \delta)\right\}; \qquad \text{(A13.1)}$$

where m_j is the expected cell frequency associated with the cell having observed frequency $x_j = \delta$. The cases $\lambda = 0$ and $\lambda = -1$ (in (6.5)) are obtained by taking the limits $\lambda \to 0$ and $\lambda \to -1$.

To prove this result, recall from (2.17) that

$$2I^\lambda(\mathbf{x} : \mathbf{m}) = \frac{2}{\lambda(\lambda+1)} \sum_{i=1}^{k} x_i\left[\left(\frac{x_i}{m_i}\right)^\lambda - 1\right]. \qquad \text{(A13.2)}$$

Differentiating (A13.2) with respect to x_i; $i = 1, \ldots, k$ it becomes clear that $2I^\lambda(\mathbf{x} : \mathbf{m})$ will be minimized when x_i/m_i is constant for all $i \neq j$. In order to satisfy the constraint that $\sum_{i=1}^k x_i = \sum_{i=1}^k m_i = n$, it follows that $x_i/m_i = (n-\delta)/(n-m_j)$ for all $i \neq j$ and $x_j/m_j = \delta/m_j$. Now we substitute these constant ratios into (A13.2) and expand in a Taylor series,

$$\begin{aligned}
2I^\lambda(\mathbf{x} : \mathbf{m}) &= \frac{2}{\lambda(\lambda+1)}\left\{\sum_{\substack{i=1\\ i\neq j}}^{k} x_i\left[\left(\frac{n-\delta}{n-m_j}\right)^\lambda - 1\right] + \delta\left[\left(\frac{\delta}{m_j}\right)^\lambda - 1\right]\right\}\\
&= \frac{2}{\lambda(\lambda+1)}\left\{(n-\delta)\left[\left(1 - \frac{m_j-\delta}{n-\delta}\right)^{-\lambda} - 1\right] + \delta\left[\left(\frac{\delta}{m_j}\right)^\lambda - 1\right]\right\}\\
&= \frac{2}{\lambda(\lambda+1)}\left\{(n-\delta)\left[\frac{\lambda(m_j-\delta)}{n-\delta} + o(n^{-1})\right] + \delta\left[\left(\frac{\delta}{m_j}\right)^\lambda - 1\right]\right\}.
\end{aligned}$$

Therefore as $n \to \infty$, we obtain (A13.1).

A14. Limiting Form of the Power-Divergence Statistic as the Parameter $\lambda \to \pm\infty$ (Section 6.2)

From (2.17)

$$2I^\lambda(\mathbf{x} : \mathbf{m}) = \frac{2}{\lambda(\lambda+1)} \sum_{i=1}^{k} x_i\left[\left(\frac{x_i}{m_i}\right)^\lambda - 1\right]; \qquad -\infty < \lambda < \infty,$$

therefore

$$[1 + \lambda(\lambda + 1)I^\lambda(\mathbf{x} : \mathbf{m})/n]^{1/\lambda} = \left[\sum_{i=1}^{k} \frac{x_i}{n}\left(\frac{x_i}{m_i}\right)^\lambda\right]^{1/\lambda}. \qquad \text{(A14.1)}$$

Assuming that the observed and expected frequency vectors $\mathbf{x}$ and $\mathbf{m}$ are not identical, it follows that $\max_{1 \le i \le k}(x_i/m_i) > 1$ (otherwise $m_i \le x_i$ for every i with strict inequality for at least one i, which contradicts $\sum_{i=1}^k m_i = \sum_{i=1}^k x_i = n$). Therefore the right-hand side of (A14.1) will converge to $\max_{1 \le i \le k}(x_i/m_i)$ as $\lambda \to \infty$, and

$$I^\lambda(\mathbf{x} : \mathbf{m})/n \sim \frac{\left[\max\limits_{1 \le i \le k} \dfrac{x_i}{m_i}\right]^\lambda - 1}{\lambda(\lambda + 1)},$$

as required in (6.8).

For $|\lambda|$ large and λ negative, the right-hand side of (A14.1) converges to $[\max_{1 \le i \le k}(m_i/x_i)]^{-1} = \min_{1 \le i \le k}(x_i/m_i)$ as $\lambda \to -\infty$. Therefore for large negative λ, (A14.1) gives

$$I^\lambda(\mathbf{x} : \mathbf{m})/n \sim \frac{\left[\min\limits_{1 \le i \le k} \dfrac{x_i}{m_i}\right]^\lambda - 1}{\lambda(\lambda + 1)}.$$

Bibliography

Agresti, A. (1984). *Analysis of Ordinal Categorical Data.* New York, John Wiley.

Agresti, A., Wackerly, D., and Boyett, J.M. (1979). Exact conditional tests for cross-classifications: approximation of attained significance levels. *Psychometrika* **44,** 75–83.

Agresti, A., and Yang, M.C. (1987). An empirical investigation of some effects of sparseness in contingency tables. *Computational Statistics and Data Analysis* **5,** 9–21.

Akaike, H. (1973). Information theory and an extension of the maximum likelihood principle. In *2nd International Symposium on Information Theory* (editors B.N. Petrov and F. Csáki), 267–281. Budapest, Akadémiai Kiadó.

Ali, S.M., and Silvey, S.D. (1966). A general class of coefficients of divergence of one distribution from another. *Journal of the Royal Statistical Society Series B* **28,** 131–142.

Altham, P.M.E. (1976). Discrete variable analysis for individuals grouped into families. *Biometrika* **63,** 263–269.

Altham, P.M.E. (1979). Detecting relationships between categorical data observed over time: a problem of deflating a χ^2 statistic. *Applied Statistics* **28,** 115–125.

Anderson, T.W. (1984). *An Introduction to Multivariate Statistical Analysis* (2nd edition). New York, John Wiley.

Anscombe, F.J. (1953). Contribution to the discussion of H. Hotelling's paper. *Journal of the Royal Statistical Society Series B* **15,** 229–230.

Anscombe, F.J. (1981). *Computing in Statistical Science through APL.* New York, Springer-Verlag.

Anscombe, F.J. (1985). Private communication.

Azencott, R., and Dacunha-Castelle, D. (1986). *Series of Irregular Observations.* New York, Springer-Verlag.

Bahadur, R.R. (1960). Stochastic comparison of tests. *Annals of Mathematical Statistics* **31,** 276–295.

Bahadur, R.R. (1965). An optimal property of the likelihood ratio statistic. *Proceedings of the Fifth Berkeley Symposium on Mathematical Statistics and Probability* **1,** 13–26.

Bahadur, R.R. (1967). Rates of convergence of estimates and test statistics. *Annals of Mathematical Statistics* **38,** 303–324.

Bahadur, R.R. (1971). *Some Limit Theorems in Statistics.* Philadelphia, Society for Industrial and Applied Mathematics.

Baker, R.J. (1977). Algorithm AS 112. Exact distributions derived from two-way tables. *Applied Statistics* **26,** 199–206.

Bednarski, T. (1980). Applications and optimality of the chi-square test of fit for testing ε-validity of parametric models. *Springer Lecture Notes in Statistics* **2** (editors W. Klonecki, A. Kozek, and J. Rosinski), 38–46.

Bednarski, T., and Ledwina, T. (1978). A note on biasedness of tests of fit. *Mathematische Operationsforschung und Statistik, Series Statistics* **9,** 191–193.

Bedrick, E.J. (1983). Adjusted chi-squared tests for cross-classified tables of survey data. *Biometrika* **70,** 591–595.

Bedrick, E.J. (1987). A family of confidence intervals for the ratio of two binomial proportions. *Biometrics* **43,** 993–998.

Bedrick, E.J., and Aragon, J. (1987). Approximate confidence intervals for the parameters of a stationary binary Markov chain. Unpublished manuscript, Department of Mathematics and Statistics, University of New Mexico, NM.

Benedetti, J.K., and Brown, M.B. (1978). Strategies for the selection of log-linear models. *Biometrics* **34,** 680–686.

Benzécri, J.P. (1973). *L'analyse des données: 1, La taxonomie; 2, L'analyse des correspondances.* Paris, Dunod.

Beran, R. (1977). Minimum Hellinger distance estimates for parametric models. *Annals of Statistics* **5,** 445–463.

Berger, T. (1983). Information theory and coding theory. In *Encyclopedia of Statistical Sciences, Volume 4* (editors S. Kotz and N.L. Johnson), 124–141. New York, John Wiley.

Berkson, J. (1978). In dispraise of the exact test. *Journal of Statistical Planning and Inference* **2,** 27–42.

Berkson, J. (1980). Minimum chi-square, not maximum likelihood! *Annals of Statistics* **8,** 457–487.

Best, D.J., and Rayner, J.C.W. (1981). Are two classes enough for the X^2 goodness of fit test? *Statistica Neerlandica* **35,** 157–163.

Best, D.J., and Rayner, J.C.W. (1985). Uniformity testing when alternatives have low order. *Sankhya Series A* **47,** 25–35.

Best, D.J., and Rayner, J.C.W. (1987). Goodness-of-fit for grouped data using components of Pearson's X^2. *Computational Statistics and Data Analysis* **5,** 53–57.

Bickel, P.J., and Rosenblatt, M. (1973). On some global measures of the deviations of density function estimates. *Annals of Statistics* **1,** 1071–1095.

Binder, D.A., Gratton, M., Hidiroglou, M.A., Kumar, S., and Rao, J.N.K. (1984). Analysis of categorical data from surveys with complex designs: some Canadian experiences. *Survey Methodology* **10,** 141–156.

Birch, M.W. (1964). A new proof of the Pearson-Fisher theorem. *Annals of Mathematical Statistics* **35,** 817–824.

Bishop, Y.M.M., Fienberg, S.E., and Holland, P.W. (1975). *Discrete Multivariate Analysis: Theory and Practice.* Cambridge, MA, The MIT Press.

Bofinger, E. (1973). Goodness-of-fit test using sample quantiles. *Journal of the Royal Statistical Society Series B* **35,** 277–284.

Böhning, D., and Holling, H. (1988). On minimizing chi-square distances under the hypothesis of homogeneity or independence for a two-way contingency table. *Statistics* (to appear).

Box, G.E.P., and Cox, D.R. (1964). An analysis of transformations. *Journal of the Royal Statistical Society Series B* **26,** 211–252.

Box, G.E.P., Hunter, W.G., and Hunter, J.S. (1978). *Statistics for Experimenters.* New York, John Wiley.

Brier, S.S. (1980). Analysis of contingency tables under cluster sampling. *Biometrika* **67,** 591–596.

Broffitt, J.D., and Randles, R.H. (1977). A power approximation for the chi-square goodness-of-fit test: simple hypothesis case. *Journal of the American Statistical Association* **72,** 604–607.

Brown, M.B. (1976). Screening effects in multidimensional contingency tables. *Applied Statistics* **25,** 37–46.

Burbea, J., and Rao, C.R. (1982). On the convexity of some divergence measures based on entropy functions. *IEEE Transactions on Information Theory* **28,** 489–495.

Burman, P. (1987). Smoothing sparse contingency tables. *Sankhya Series A* **49,** 24–36.

Causey, B.D. (1983). Estimation of proportions for multinomial contingency tables subject to marginal constraints. *Communications in Statistics—Theory and Methods* **12,** 2581–2587.

Chandra, T.K., and Ghosh, J.K. (1980). Valid asymptotic expansions for the likelihood ratio and other statistics under contiguous alternatives. *Sankhya Series A* **42,** 170–184.

Chapman, J.W. (1976). A comparison of the X^2, $-2\log R$, and multinomial probability criteria for significance tests when expected frequencies are small. *Journal of the American Statistical Association* **71,** 854–863.

Chernoff, H., and Lehmann, E.L. (1954). The use of maximum likelihood estimates in χ^2 tests for goodness of fit. *Annals of Mathematical Statistics* **25,** 579–586.

Cleveland, W.S. (1985). *The Elements of Graphing Data.* Monterey, CA, Wadsworth.

Cochran, W.G. (1942). The χ^2 correction for continuity. *Iowa State College Journal of Science* **16,** 421–436.

Cochran, W.G. (1952). The χ^2 test of goodness of fit. *Annals of Mathematical Statistics* **23,** 315–345.

Cochran, W.G. (1954). Some methods for strengthening the common χ^2 tests. *Biometrics* **10,** 417–451.

Cochran, W.G., and Cox, G.M. (1957). *Experimental Designs* (2nd edition). New York, John Wiley.

Cohen, A., and Sackrowitz, H.B. (1975). Unbiasedness of the chi-square, likelihood ratio, and other goodness of fit tests for the equal cell case. *Annals of Statistics* **3,** 959–964.

Cohen, J.E. (1976). The distribution of the chi-squared statistic under cluster sampling from contingency tables. *Journal of the American Statistical Association* **71,** 665–670.

Cox, C. (1984). An elementary introduction to maximum likelihood estimation for multinomial models: Birch's theorem and the delta method. *American Statistician* **38,** 283–287.

Cox, D.R. (1970). *The Analysis of Binary Data.* London, Methuen.

Cox, D.R., and Hinkley, D.V. (1974). *Theoretical Statistics.* London, Chapman and Hall.

Cox, M.A.A., and Plackett, R.L. (1980). Small samples in contingency tables. *Biometrika* **67,** 1–13.

Cramér, H. (1946). *Mathematical Methods of Statistics.* Princeton, NJ, Princeton University Press.

Cressie, N. (1976). On the logarithms of high-order spacings. *Biometrika* **63,** 343–355.

Cressie, N. (1979). An optimal statistic based on higher order gaps. *Biometrika* **66,** 619–627.

Cressie, N. (1988). Estimating census undercount at national and subnational levels. In *Proceedings of the Bureau of the Census 4th Annual Research Conference.* Washington, DC, U.S. Bureau of the Census.

Cressie, N., and Read, T.R.C. (1984). Multinomial goodness-of-fit tests. *Journal of the Royal Statistical Society Series B* **46,** 440–464.

Cressie, N., and Read, T.R.C. (1988). Cressie-Read statistic. In *Encyclopedia of Statistical Sciences, Supplementary Volume* (editors S. Kotz and N.L. Johnson) (to appear). New York, John Wiley.

Csáki, E., and Vince, I. (1977). On modified forms of Pearson's chi-square. *Bulletin of the International Statistical Institute* **47,** 669–672.
Csáki, E., and Vince, I. (1978). On limiting distribution laws of statistics analogous to Pearson's chi-square. *Mathematische Operationsforschung und Statistik, Series Statistics* **9,** 531–548.
Csiszár, I. (1978). Information measures: a critical survey. In *Transactions of the Seventh Prague Conference on Information Theory, Statistical Decision Functions and Random Processes and of the 1974 European Meeting of Statisticians, Volume B,* 73–86. Dordrecht, Reidel.
Dahiya, R.C., and Gurland, J. (1972). Pearson chi-square test of fit with random intervals. *Biometrika* **59,** 147–153.
Dahiya, R.C., and Gurland, J. (1973). How many classes in the Pearson chi-square test? *Journal of the American Statistical Association* **68,** 707–712.
Dale, J.R. (1986). Asymptotic normality of goodness-of-fit statistics for sparse product multinomials. *Journal of the Royal Statistical Society Series B* **48,** 48–59.
Darling, D.A. (1953). On a class of problems related to the random division of an interval. *Annals of Mathematical Statistics* **24,** 239–253.
Darroch, J.N. (1974). Multiplicative and additive interactions in contingency tables. *Biometrika* **61,** 207–214.
Darroch, J.N. (1976). No-interaction in contingency tables. In *Proceedings of the 9th International Biometric Conference*, 296–316. Raleigh, NC, The Biometric Society.
Darroch, J.N., and Speed, T.P. (1983). Additive and multiplicative models and interactions. *Annals of Statistics* **11,** 724–738.
del Pino, G.E. (1979). On the asymptotic distribution of k-spacings with applications to goodness-of-fit tests. *Annals of Statistics* **7,** 1058–1065.
Denteneer, D., and Verbeek, A. (1986). A fast algorithm for iterative proportional fitting in log-linear models. *Computational Statistics and Data Analysis* **3,** 251–264.
Dillon, W.R., and Goldstein, M. (1984). *Multivariate Analysis: Methods and Applications.* New York, John Wiley.
Draper, N.R., and Smith, H. (1981). *Applied Regression Analysis* (2nd edition). New York, John Wiley.
Drost, F.C., Kallenberg, W.C.M., Moore, D.S., and Oosterhoff, J. (1987). Power approximations to multinomial tests of fit. *Memorandum Nr. 633*, Faculty of Applied Mathematics, University of Twente, The Netherlands.
Dudewicz, E.J., and van der Meulen, E.C. (1981). Entropy-based tests of uniformity. *Journal of the American Statistical Association* **76,** 967–974.
Durbin, J. (1978). Goodness-of-fit tests based on the order statistics. In *Transactions of the Seventh Prague Conference on Information Theory, Statistical Decision Functions and Random Processes and of the 1974 European Meeting of Statisticians, Volume B*, 109–118. Dordrecht, Reidel.
Fay, R.E. (1985). A jackknifed chi-squared test for complex samples. *Journal of the American Statistical Association* **80,** 148–157.
Ferguson, T.S. (1967). *Mathematical Statistics: A Decision Theoretic Approach.* New York, Academic Press.
Fienberg, S.E. (1979). The use of chi-squared statistics for categorical data problems. *Journal of the Royal Statistical Society Series B* **41,** 54–64.
Fienberg, S.E. (1980). *The Analysis of Cross-Classified Categorical Data* (2nd edition). Cambridge, MA, The MIT Press.
Fienberg, S.E. (1984). The contributions of William Cochran to categorical data analysis. In *W.G. Cochran's Impact on Statistics* (editors P.S.R.S. Rao and J. Sedransk), 103–118. New York, John Wiley.
Fienberg, S.E. (1986). Adjusting the census: statistical methodology for going beyond the count. *Proceedings of the 2nd Annual Research Conference*, 570–577. Washington, DC, U.S. Bureau of the Census.
Fisher, R.A. (1924). The conditions under which χ^2 measures the discrepancy between

observation and hypothesis. *Journal of the Royal Statistical Society* **87,** 442–450.

Forthofer, R.N., and Lehnen, R.G. (1981). *Public Program Analysis: A New Categorical Data Approach.* Belmont, CA, Lifetime Learning Publications.

Freeman, D.H. (1987). *Applied Categorical Data Analysis.* New York, Marcel Dekker.

Freeman, M.F., and Tukey, J.W. (1950). Transformations related to the angular and the square root. *Annals of Mathematical Statistics* **21,** 607–611.

Frosini, B.V. (1976). On the power function of the χ^2 test. *Metron* **34,** 3–36.

Gan, F.F. (1985). Goodness-of-fit statistics for location-scale distributions. Ph.D. Dissertation, Department of Statistics, Iowa State University, Ames, IA.

Gebert, J.R. (1968). A power study of Kimball's statistics. *Statistische Hefte* **9,** 269–273.

Gebert, J.R., and Kale, B.K. (1969). Goodness of fit tests based on discriminatory information. *Statistiche Hefte* **10,** 192–200.

Gibbons, J.D., and Pratt, J.W. (1975). *P*-values: interpretation and methodology. *American Statistician* **29,** 20–25.

Gleser, L.J., and Moore, D.S. (1985). The effect of positive dependence on chi-squared tests for categorical data. *Journal of the Royal Statistical Society Series B* **47,** 459–465.

Gokhale, D.V., and Kullback, S. (1978). *The Information in Contingency Tables.* New York, Marcel Dekker.

Goldstein, M., Wolf, E., and Dillon, W. (1976). On a test of independence for contingency tables. *Communications in Statistics—Theory and Methods* **5,** 159–169.

Good, I.J., Gover, T.N., and Mitchell, G.J. (1970). Exact distributions for X^2 and for the likelihood-ratio statistic for the equiprobable multinomial distribution. *Journal of the American Statistical Association* **65,** 267–283.

Goodman, L.A. (1984). *Analysis of Cross-Classified Data Having Ordered Categories.* Cambridge, MA, Harvard University Press.

Goodman, L.A. (1986). Some useful extensions of the usual correspondence analysis approach and the usual log-linear models approach in the analysis of contingency tables. *International Statistical Review* **54,** 243–270.

Greenwood, M. (1946). The statistical study of infectious diseases. *Journal of the Royal Statistical Society Series A* **109,** 85–110.

Grizzle, J.E., Starmer, C.F., and Koch, G.G. (1969). Analysis of categorical data by linear models. *Biometrics* **25,** 489–504.

Grove, D.M. (1986). Positive association in a two-way contingency table: a numerical study. *Communications in Statistics—Simulation and Computation* **15,** 633–648.

Gumbel, E.J. (1943). On the reliability of the classical chi-square test. *Annals of Mathematical Statistics* **14,** 253–263.

Guttorp, P., and Lockhart, R.A. (1988). On the asymptotic distribution of quadratic forms in uniform order statistics. *Annals of Statistics* **16,** 433–449.

Gvanceladze, L.G., and Chibisov, D.M. (1979). On tests of fit based on grouped data. In *Contributions to Statistics, J. Hájek Memorial Volume* (editor J. Jurečková), 79–89. Prague, Academia.

Haber, M. (1980). A comparative simulation study of the small sample powers of several goodness of fit tests. *Journal of Statistical Computation and Simulation* **11,** 241–250.

Haber, M. (1984). A comparison of tests for the hypothesis of no three-factor interaction in 2 × 2 × 2 contingency tables. *Journal of Statistical Computation and Simulation* **20,** 205–215.

Haberman, S.J. (1974). *The Analysis of Frequency Data.* Chicago, University of Chicago Press.

Haberman, S.J. (1977). Log-linear models and frequency tables with small expected cell counts. *Annals of Statistics* **5,** 1148–1169.

Haberman, S.J. (1978). *Analysis of Qualitative Data, Volume 1.* New York, Academic Press.

Haberman, S.J. (1979). *Analysis of Qualitative Data, Volume 2.* New York, Academic Press.

Haberman, S.J. (1982). Analysis of dispersion of multinomial responses. *Journal of the American Statistical Association* **77,** 568–580.

Haberman, S.J. (1986). A warning on the use of chi-square statistics with frequency tables with small expected cell counts. Unpublished manuscript, Department of Statistics, Northwestern University, Evanston, IL.

Haldane, J.B.S. (1937). The exact value of the moments of the distribution of χ^2, used as a test of goodness of fit, when expectations are small. *Biometrika* **29,** 133–143.

Haldane, J.B.S. (1939). The mean and variance of χ^2, when used as a test of homogeneity, when expectations are small. *Biometrika* **31,** 346–355.

Hall, P. (1985). Tailor-made tests of goodness of fit. *Journal of the Royal Statistical Society Series B* **47,** 125–131.

Hall, P. (1986). On powerful distributional tests based on sample spacings. *Journal of Multivariate Analysis* **19,** 201–224.

Hamdan, M. (1963). The number and width of classes in the chi-square test. *Journal of the American Statistical Association* **58,** 678–689.

Hancock, T.W. (1975). Remark on algorithm 434 [G2]. Exact probabilities for $r \times c$ contingency tables. *Communications of the ACM* **18,** 117–119.

Harris, R.R., and Kanji, G.K. (1983). On the use of minimum chi-square estimation. *The Statistician* **32,** 379–394.

Harrison, R.H. (1985). Choosing the optimum number of classes in the chi-square test for arbitrary power levels. *Sankhya Series B* **47,** 319–324.

Havrda, J., and Charvát, F. (1967). Quantification method of classification processes: concept of structural α-entropy. *Kybernetica* **3,** 30–35.

Hayakawa, T. (1977). The likelihood ratio criterion and the asymptotic expansion of its distribution. *Annals of the Institute of Statistical Mathematics* **29,** 359–378.

Haynam, G.E., and Leone, F.C. (1965). Analysis of categorical data. *Biometrika* **52,** 654–660.

Hill, M.O. (1973). Diversity and evenness: a unifying notation and its consequences. *Ecology* **54,** 427–432.

Hodges, J.L., and Lehmann, E.L. (1970). Deficiency. *Annals of Mathematical Statistics* **41,** 783–801.

Hoeffding, W. (1965). Asymptotically optimal tests for multinomial distributions. *Annals of Mathematical Statistics* **36,** 369–408.

Hoel, P.G. (1938). On the chi-square distribution for small samples. *Annals of Mathematical Statistics* **9,** 158–165.

Holst, L. (1972). Asymptotic normality and efficiency for certain goodness-of-fit tests. *Biometrika* **59,** 137–145, 699.

Holst, L., and Rao, J.S. (1981). Asymptotic spacings theory with applications to the two-sample problem. *Canadian Journal of Statistics* **9,** 79–89.

Holt, D., Scott, A.J., and Ewings, P.D. (1980). Chi-squared tests with survey data. *Journal of the Royal Statistical Society Series A* **143,** 303–320.

Holtzman, G.I., and Good, I.J. (1986). The Poisson and chi-squared approximations as compared with the true upper-tail probability of Pearson's X^2 for equiprobable multinomials. *Journal of Statistical Planning and Inference* **13,** 283–295.

Horn, S.D. (1977). Goodness-of-fit tests for discrete data: a review and an application to a health impairment scale. *Biometrics* **33,** 237–248.

Hosmane, B. (1986). Improved likelihood ratio tests and Pearson chi-square tests for independence in two dimensional contingency tables. *Communications in Statistics—Theory and Methods* **15,** 1875–1888.

Hosmane, B. (1987). An empirical investigation of chi-square tests for the hypothesis of no three-factor interaction in $I \times J \times K$ contingency tables. *Journal of Statistical Computation and Simulation* **28,** 167–178.

Hutchinson, T.P. (1979). The validity of the chi-squared test when expected frequencies are small: a list of recent research references. *Communications in Statistics—Theory and Methods* **8,** 327–335.

Ivchenko, G.I., and Medvedev, Y.I. (1978). Separable statistics and hypothesis testing. The case of small samples. *Theory of Probability and Its Applications* **23,** 764–775.

Ivchenko, G.I., and Medvedev, Y.I. (1980). Decomposable statistics and hypothesis testing for grouped data. *Theory of Probability and Its Applications* **25,** 540–551.

Ivchenko, G.I., and Tsukanov, S.V. (1984). On a new way of treating frequencies in the method of grouping observations, and the optimality of the χ^2 test. *Soviet Mathematics Doklady* **30,** 79–82.

Jammalamadaka, S.R., and Tiwari, R.C. (1985). Asymptotic comparisons of three tests for goodness of fit. *Journal of Statistical Planning and Inference* **12,** 295–304.

Jammalamadaka, S.R., and Tiwari, R.C. (1987). Efficiencies of some disjoint spacings tests relative to a χ^2 test. In *New Perspectives in Theoretical and Applied Statistics* (editors M.L. Puri, J.P. Vilaplana, and W. Wertz), 311–317. New York, John Wiley.

Jeffreys, H. (1948). *Theory of Probability* (2nd edition). London, Oxford University Press.

Jensen, D.R. (1973). Monotone bounds on the chi-square approximation to the distribution of Pearson's X^2 statistic. *Australian Journal of Statistics* **15,** 65–70.

Johnson, N.L., and Kotz, S. (1969). *Distributions in Statistics: Discrete Distributions.* Boston, Houghton Mifflin.

Kale, B.K., and Godambe, V.P. (1967). A test of goodness of fit. *Statistische Hefte* **8,** 165–172.

Kallenberg, W.C.M. (1985). On moderate and large deviations in multinomial distributions. *Annals of Statistics* **13,** 1554–1580.

Kallenberg, W.C.M., Oosterhoff, J., and Schriever, B.F. (1985). The number of classes in chi-squared goodness-of-fit tests. *Journal of the American Statistical Association* **80,** 959–968.

Kannappan, P., and Rathie, P.N. (1978). On a generalized directed divergence and related measures. In *Transactions of the Seventh Prague Conference on Information Theory, Statistical Decision Functions and Random Processes and of the 1974 European Meeting of Statisticians, Volume B,* 255–265. Dordrecht, Reidel.

Kempton, R.A. (1979). The structure of species abundance and measurement of diversity. *Biometrics* **35,** 307–321.

Kendall, M., and Stuart, A. (1969). *The Advanced Theory of Statistics, Volume 1* (3rd edition). London, Griffin.

Kihlberg, J.K., Narragon, E.A., and Campbell, B.J. (1964). Automobile crash injury in relation to car size. *Cornell Aerospace Laboratory Report No. VJ-1823-R11,* Cornell University, Ithaca, NY.

Kimball, B.F. (1947). Some basic theorems for developing tests of fit for the case of nonparametric probability distribution functions. *Annals of Mathematical Statistics* **18,** 540–548.

Kirmani, S.N.U.A. (1973). On a goodness of fit test based on Matusita's measure of distance. *Annals of the Institute of Statistical Mathematics* **25,** 493–500.

Kirmani, S.N.U.A., and Alam, S.N. (1974). On goodness of fit tests based on spacings. *Sankhya Series A* **36,** 197–203.

Koehler, K.J. (1977). Goodness-of-fit statistics for large sparse multinomials. Ph.D. Dissertation, School of Statistics, University of Minnesota, Minneapolis, MN.

Koehler, K.J. (1986). Goodness-of-fit tests for log-linear models in sparse contingency tables. *Journal of the American Statistical Association* **81,** 483–493.

Koehler, K.J., and Larntz, K. (1980). An empirical investigation of goodness-of-fit statistics for sparse multinomials. *Journal of the American Statistical Association* **75,** 336–344.

Koehler, K.J., and Wilson, J.R. (1986). Chi-square tests for comparing vectors of proportions for several cluster samples. *Communications in Statistics—Theory and Methods* **15,** 2977–2990.

Kotze, T.J.v.W., and Gokhale, D.V. (1980). A comparison of the Pearson-X^2 and the log-likelihood-ratio statistics for small samples by means of probability ordering. *Journal of Statistical Computation and Simulation* **12,** 1–13.

Kullback, S. (1959). *Information Theory and Statistics.* New York, John Wiley.

Kullback, S. (1983). Kullback information. In *Encyclopedia of Statistical Sciences, Volume 4* (editors S. Kotz and N.L. Johnson), 421–425. New York, John Wiley.

Kullback, S. (1985). Minimum discrimination information (MDI) estimation. In *Encyclopedia of Statistical Sciences, Volume 5* (editors S. Kotz and N.L. Johnson), 527–529. New York, John Wiley.

Kullback, S., and Keegel, J.C. (1984). Categorical data problems using information theoretic approach. In *Handbook of Statistics, Volume 4* (editors P.R. Krishnaiah and P.K. Sen), 831–871. New York, Elsevier Science Publishers.

Lancaster, H.O. (1969). *The Chi-Squared Distribution.* New York, John Wiley.

Lancaster, H.O. (1980). Orthogonal models in contingency tables. In *Developments in Statistics, Volume 3* (editor P.R. Krishnaiah), Chapter 2. New York, Academic Press.

Lancaster, H.O., and Brown, T.A.I. (1965). Sizes of the chi-square test in the symmetric multinomial. *Australian Journal of Statistics* **7,** 40–44.

Larntz, K. (1978). Small-sample comparisons of exact levels for chi-squared goodness-of-fit statistics. *Journal of the American Statistical Association* **73,** 253–263.

Lau, K. (1985). Characterization of Rao's quadratic entropies. *Sankhya Series A* **47,** 295–309.

Lawal, H.B. (1980). Tables of percentage points of Pearson's goodness-of-fit statistic for use with small expectations. *Applied Statistics* **29,** 292–298.

Lawal, H.B. (1984). Comparisons of X^2, Y^2, Freeman-Tukey and Williams's improved G^2 test statistics in small samples of one-way multinomials. *Biometrika* **71,** 415–458.

Lawal, H.B., and Upton, G.J.G. (1980). An approximation to the distribution of the X^2 goodness-of-fit statistic for use with small expectations. *Biometrika* **67,** 447–453.

Lawal, H.B., and Upton, G.J.G. (1984). On the use of X^2 as a test of independence in contingency tables with small cell expectations. *Australian Journal of Statistics* **26,** 75–85.

Lawley, D.N. (1956). A general method for approximating to the distribution of likelihood ratio criteria. *Biometrika* **43,** 295–303.

Lee, C.C. (1987). Chi-squared tests for and against an order restriction on multinomial parameters. *Journal of the American Statistical Association* **82,** 611–618.

Levin, B. (1983). On calculations involving the maximum cell frequency. *Communications in Statistics—Theory and Methods* **12,** 1299–1327.

Lewis, P.A.W., Liu, L.H., Robinson, D.W., and Rosenblatt, M. (1977). Empirical sampling study of a goodness of fit statistic for density function estimation. In *Multivariate Analysis, Volume 4* (editor P.R. Krishnaiah), 159–174. Amsterdam, North-Holland.

Lewis, T., Saunders, I.W., and Westcott, M. (1984). The moments of the Pearson chi-squared statistic and the minimum expected value in two-way tables. *Biometrika* **71,** 515–522.

Lewontin, R., and Felsenstein, J. (1965). The robustness of homogeneity tests in $2 \times n$ tables. *Biometrics* **21,** 19–33.

Makridakis, S., Andersen, A., Carbone, R., Fildes, R., Hibon, M., Lewandowski, R., Newton, J., Parzen, E., and Winkler, R. (1982). The accuracy of extrapolation (time series) methods: results of a forecasting competition. *Journal of Forecasting* **1,** 111–153.

Mann, H.B., and Wald, A. (1942). On the choice of the number of class intervals in the application of the chi-square test. *Annals of Mathematical Statistics* **13,** 306–317.

Margolin, B.H., and Light, R.L. (1974). An analysis of variance for categorical data,

II: small sample comparisons with chi square and other competitors. *Journal of the American Statistical Association* **69,** 755–764.

Mathai, A.M., and Rathie, P.N. (1972). Characterization of Matusita's measure of affinity. *Annals of the Institute of Statistical Mathematics* **24,** 473–482.

Mathai, A.M., and Rathie, P.N. (1975). *Basic Concepts in Information Theory and Statistics.* New York, John Wiley.

Mathai, A.M., and Rathie, P.N. (1976). Recent contributions to axiomatic definitions of information and statistical measures through functional equations. In *Essays in Probability and Statistics—A Volume in Honor of Professor Junjiro Ogawa* (editors S. Ikeda, T. Hayakawa, H. Hudimoto, M. Okamoto, M. Siotani, and S. Yamamoto), 607–633. Tokyo, Shinko Tsusho.

Matusita, K. (1954). On the estimation by the minimum distance method. *Annals of the Institute of Statistical Mathematics* **5,** 59–65.

Matusita, K. (1955). Decision rules based on the distance, for problems of fit, two samples, and estimation. *Annals of Mathematical Statistics* **26,** 631–640.

Matusita, K. (1971). Some properties of affinity and applications. *Annals of the Institute of Statistical Mathematics* **23,** 137–156.

McCullagh, P. (1985a). On the asymptotic distribution of Pearson's statistic in linear exponential-family models. *International Statistical Review* **53,** 61–67.

McCullagh, P. (1985b). Sparse data and conditional tests. *Bulletin of the International Statistical Institute, Proceedings of the 45th Session—Amsterdam* **51,** 28.3-1–28.3-10.

McCullagh, P. (1986). The conditional distribution of goodness-of-fit statistics for discrete data. *Journal of the American Statistical Association* **81,** 104–107.

McCullagh, P., and Nelder, J.A. (1983). *Generalized Linear Models.* London, Chapman and Hall.

Mehta, C.R., and Patel, N.R. (1986). Algorithm 643. FEXACT: A FORTRAN subroutine for Fisher's exact test on unordered $r \times c$ contingency tables. *ACM Transactions on Mathematical Software* **12,** 154–161.

Mickey, R.M. (1987). Assessment of three way interaction in $2 \times J \times K$ tables. *Computational Statistics and Data Analysis* **5,** 23–30.

Mitra, S.K. (1958). On the limiting power function of the frequency chi-square test. *Annals of Mathematical Statistics* **29,** 1221–1233.

Miyamoto, Y. (1976). Optimum spacing for goodness of fit test based on sample quantiles. In *Essays in Probability and Statistics—A Volume in Honor of Professor Junjiro Ogawa* (editors S. Ikeda, T. Hayakawa, H. Hudimoto, M. Okamoto, M. Siotani, and S. Yamamoto), 475–483. Tokyo, Shinko Tsusho.

Moore, D.S. (1986). Tests of chi-squared type. In *Goodness-of-Fit Techniques* (editors R.B. D'Agostino and M.A. Stephens), 63–95. New York, Marcel Dekker.

Moore, D.S., and Spruill, M.C. (1975). Unified large-sample theory of general chi-squared statistics for tests of fit. *Annals of Statistics* **3,** 599–616.

Morris, C. (1966). Admissible Bayes procedures and classes of epsilon Bayes procedures for testing hypotheses in a multinomial distribution. *Technical Report 55*, Department of Statistics, Stanford University, Stanford, CA.

Morris, C. (1975) Central limit theorems for multinomial sums. *Annals of Statistics* **3,** 165–188.

Muirhead, R.J. (1982). *Aspects of Multivariate Statistical Theory.* New York, John Wiley.

Nath, P. (1972). Some axiomatic characterizations of a non-additive measure of divergence in information. *Journal of Mathematical Sciences* **7,** 57–68.

National Center for Health Statistics (1970). *Vital Statistics of the United States, 1970,* **2,** *Part A*. Washington, D.C., U.S. Government Printing Office.

Nayak, T.K. (1985). On diversity measures based on entropy functions. *Communications in Statistics—Theory and Methods* **14,** 203–215.

Nayak, T.K. (1986). Sampling distributions in analysis of diversity. *Sankhya Series B* **48,** 1–9.

Neyman, J. (1949). Contribution to the theory of the χ^2 test. *Proceedings of the First Berkeley Symposium on Mathematical Statistics and Probability*, 239–273.
Oosterhoff, J. (1985). The choice of cells in chi-square tests. *Statistica Neerlandica* **39,** 115–128.
Oosterhoff, J., and Van Zwet, W.R. (1972). The likelihood ratio test for the multinomial distribution. *Proceedings of the Sixth Berkeley Symposium on Mathematical Statistics and Probability* **2,** 31–49.
Parr, W.C. (1981). Minimum distance estimation: a bibliography. *Communications in Statistics—Theory and Methods* **10,** 1205–1224.
Patil, G.P., and Taillie, C. (1982). Diversity as a concept and its measurement. *Journal of the American Statistical Association* **77,** 548–561.
Patni, G.C., and Jain, K.C. (1977). On axiomatic characterization of some non-additive measures of information. *Metrika* **24,** 23–34.
Pearson, K. (1900). On the criterion that a given system of deviations from the probable in the case of a correlated system of variables is such that it can be reasonably supposed to have arisen from random sampling. *Philosophy Magazine* **50,** 157–172.
Peers, H.W. (1971). Likelihood ratio and associated test criteria. *Biometrika* **58,** 577–587.
Pettitt, A.N., and Stephens, M.A. (1977). The Kolmogorov-Smirnov goodness-of-fit statistic with discrete and grouped data. *Technometrics* **19,** 205–210.
Plackett, R.L. (1981). *The Analysis of Categorical Data* (2nd edition). High Wycombe, Griffin.
Pollard, D. (1979). General chi-square goodness-of-fit tests with data-dependent cells. *Zeitschrift fur Wahrscheinlichkeitstheorie und verwandte Gebiete* **50,** 317–331.
Ponnapalli, R. (1976). Deficiencies of minimum discrepancy estimators. *Canadian Journal of Statistics* **4,** 33–50.
Pyke, R. (1965). Spacings. *Journal of the Royal Statistical Society Series B* **27,** 395–449.
Quade, D., and Salama, I.A. (1975). A note on minimum chi-square statistics in contingency tables. *Biometrics* **31,** 953–956.
Quine, M.P., and Robinson, J. (1985). Efficiencies of chi-square and likelihood ratio goodness-of-fit tests. *Annals of Statistics* **13,** 727–742.
Radlow, R., and Alf, E.F. (1975). An alternate multinomial assessment of the accuracy of the χ^2 test of goodness of fit. *Journal of the American Statistical Association* **70,** 811–813.
Raftery, A.E. (1986a). Choosing models for cross-classifications. *American Sociological Review* **51,** 145–146.
Raftery, A.E. (1986b). A note on Bayes factors for log-linear contingency table models with vague prior information. *Journal of the Royal Statistical Society Series B* **48,** 249–250.
Rao, C.R. (1961). Asymptotic efficiency and limiting information. *Proceedings of the Fourth Berkeley Symposium on Mathematical Statistics and Probability* **1,** 531–546.
Rao, C.R. (1962). Efficient estimates and optimum inference procedures in large samples. *Journal of the Royal Statistical Society Series B* **24,** 46–72.
Rao, C.R. (1963). Criteria of estimation in large samples. *Sankhya Series A* **25,** 189–206.
Rao, C.R. (1973). *Linear Statistical Inference and Its Applications* (2nd edition). New York, John Wiley.
Rao, C.R. (1982a). Diversity and dissimilarity coefficients: a unified approach. *Theoretical Population Biology* **21,** 24–43.
Rao, C.R. (1982b). Diversity: its measurement, decomposition, apportionment and analysis. *Sankhya Series A* **44,** 1–22.
Rao, C.R. (1986). ANODIV: generalization of ANOVA through entropy and cross entropy functions (presented at 4th Vilnius Conference, Vilnius, USSR, 1985). In *Probability Theory and Mathematical Statistics, Volume 2* (editors Y.V. Prohorov, V.A. Statulevicius, V.V. Sazonov, and B. Grigelionis), 477–494. Utrecht, VNU Science Press.

Rao, C.R., and Nayak, T.K. (1985). Cross entropy, dissimilarity measures, and characterizations of quadratic entropy. *IEEE Transactions on Information Theory* **31,** 589–593.

Rao, J.N.K., and Scott, A.J. (1981). The analysis of categorical data from complex sample surveys: chi-squared tests for goodness of fit and independence in two-way tables. *Journal of the American Statistical Association* **76,** 221–230.

Rao, J.N.K., and Scott, A.J. (1984). On chi-squared tests for multiway contingency tables with cell proportions estimated from survey data. *Annals of Statistics* **12,** 46–60.

Rao, J.S., and Kuo, M. (1984). Asymptotic results on the Greenwood statistic and some of its generalizations. *Journal of the Royal Statistical Society Series B* **46,** 228–237.

Rao, K.C., and Robson, D.S. (1974). A chi-square statistic for goodness-of-fit tests within the exponential family. *Communications in Statistics—Theory and Methods* **3,** 1139–1153.

Rathie, P.N. (1973). Some characterization theorems for generalized measures of uncertainty and information. *Metrika* **20,** 122–130.

Rathie, P.N., and Kannappan, P. (1972). A directed-divergence function of type β. *Information and Control* **20,** 38–45.

Rayner, J.C.W., and Best, D.J. (1982). The choice of class probabilities and number of classes for the simple X^2 goodness of fit test. *Sankhya Series B* **44,** 28–38.

Rayner, J.C.W., Best, D.J., and Dodds, K.G. (1985). The construction of the simple X^2 and Neyman smooth goodness of fit tests. *Statistica Neerlandica* **39,** 35–50.

Read, T.R.C. (1982). Choosing a goodness-of-fit test. Ph.D. Dissertation, School of Mathematical Sciences, The Flinders University of South Australia, Adelaide, South Australia.

Read, T.R.C. (1984a). Closer asymptotic approximations for the distributions of the power divergence goodness-of-fit statistics. *Annals of the Institute of Statistical Mathematics* **36,** 59–69.

Read, T.R.C. (1984b). Small-sample comparisons for the power divergence goodness-of-fit statistics. *Journal of the American Statistical Association* **79,** 929–935.

Read, T.R.C., and Cowan, R. (1976). Probabilistic modelling and hypothesis testing applied to permutation data. Private correspondence.

Rényi, A. (1961). On measures of entropy and information. *Proceedings of the Fourth Berkeley Symposium on Mathematical Statistics and Probability* **1,** 547–561.

Roberts, G., Rao, J.N.K., and Kumar, S. (1987). Logistic regression analysis of sample survey data. *Biometrika* **74,** 1–12.

Roscoe, J.T., and Byars, J.A. (1971). An investigation of the restraints with respect to sample size commonly imposed on the use of the chi-square statistic. *Journal of the American Statistical Association* **66,** 755–759.

Roy, A.R. (1956). On χ^2 statistics with variable intervals. *Technical Report 1*, Department of Statistics, Stanford University, Stanford, CA.

Rudas, T. (1986). A Monte Carlo comparison of the small sample behaviour of the Pearson, the likelihood ratio and the Cressie-Read statistics. *Journal of Statistical Computation and Simulation* **24,** 107–120.

Sakamoto, Y., and Akaike, H. (1978). Analysis of cross classified data by AIC. *Annals of the Institute of Statistical Mathematics Part B* **30,** 185–197.

Sakamoto, Y., Ishiguro, M., and Kitagawa, G. (1986). *Akaike Information Criterion Statistics.* Tokyo, KTK Scientific Publishers.

Sakhanenko, A.I., and Mosyagin, V.E. (1977). Speed of convergence of the distribution of the likelihood ratio statistic. *Siberian Mathematical Journal* **18,** 1168–1175.

SAS Institute Inc. (1982). *SAS User's Guide: Statistics.* Cary, NC, SAS Institute.

Schwarz, G. (1978). Estimating the dimension of a model. *Annals of Statistics* **6,** 461–464.

Selivanov, B.I. (1984). Limit distributions of the χ^2 statistic. *Theory of Probability and Its Applications* **29,** 133–134.

Sharma, B.D., and Taneja, I.J. (1975). Entropy of type (α, β) and other generalized measures in information theory. *Metrika* **22,** 205–215.

Silverman, B.W. (1982). On the estimation of a probability density function by the maximum penalized likelihood method. *Annals of Statistics* **10,** 795–810.

Simonoff, J.S. (1983). A penalty function approach to smoothing large sparse contingency tables. *Annals of Statistics* **11,** 208–218.

Simonoff, J.S. (1985). An improved goodness-of-fit statistic for sparse multinomials. *Journal of the American Statistical Association* **80,** 671–677.

Simonoff, J.S. (1986). Jackknifing and bootstrapping goodness-of-fit statistics in sparse multinomials. *Journal of the American Statistical Association* **81,** 1005–1011.

Simonoff, J.S. (1987). Probability estimation via smoothing in sparse contingency tables with ordered categories. *Statistics and Probability Letters* **5,** 55–63.

Sinha, B.K. (1976). On unbiasedness of Mann-Wald-Gumbell χ^2 test. *Sankhya Series A* **38,** 124–130.

Siotani, M., and Fujikoshi, Y. (1984). Asymptotic approximations for the distributions of multinomial goodness-of-fit statistics. *Hiroshima Mathematics Journal* **14,** 115–124.

Slakter, M.J. (1966). Comparative validity of the chi-square and two modified chi-square goodness of fit tests for small but equal expected frequencies. *Biometrika* **53,** 619–622.

Slakter, M.J. (1968). Accuracy of an approximation to the power function of the chi-square goodness of fit test with small but equal expected frequencies. *Journal of the American Statistical Association* **63,** 912–918.

Smith, P.J., Rae, D.S., Manderscheid, R.W., and Silbergeld, S. (1979). Exact and approximate distributions of the chi-square statistic for equiprobability. *Communications in Statistics—Simulation and Computation* **8,** 131–149.

Smith, P.J., Rae, D.S., Manderscheid, R.W., and Silbergeld, S. (1981). Approximating the moments and distribution of the likelihood ratio statistic for multinomial goodness of fit. *Journal of the American Statistical Association* **76,** 737–740.

Spencer, B. (1985). Statistical aspects of equitable apportionment. *Journal of the American Statistical Association* **80,** 815–822.

Spruill, C. (1977). Equally likely intervals in the chi-square test. *Sankhya Series A* **39,** 299–302.

Steck, G. (1957). Limit theorems for conditional distributions. *University of California Publications in Statistics* **2,** 237–284.

Stephens, M.A. (1986a). Tests based on EDF statistics. In *Goodness-of-Fit Techniques* (editors R.B. D'Agostino and M.A. Stephens), 97–193. New York, Marcel Dekker.

Stephens, M.A. (1986b). Tests for the uniform distribution. In *Goodness-of-Fit Techniques* (editors R.B. D'Agostino and M.A. Stephens), 331–366. New York, Marcel Dekker.

Stevens, S.S. (1975). *Psychophysics.* New York, John Wiley.

Sutrick, K.H. (1986). Asymptotic power comparisons of the chi-square and likelihood ratio tests. *Annals of the Institute of Statistical Mathematics* **38,** 503–511.

Tate, M.W., and Hyer, L.A. (1973). Inaccuracy of the X^2 test of goodness of fit when expected frequencies are small. *Journal of the American Statistical Association* **68,** 836–841.

Tavaré, S. (1983). Serial dependence in contingency tables. *Journal of the Royal Statistical Society Series B* **45,** 100–106.

Tavaré, S., and Altham, P.M.E. (1983). Serial dependence of observations leading to contingency tables, and corrections to chi-squared statistics. *Biometrika* **70,** 139–144.

Thomas, D.R., and Rao, J.N.K. (1987). Small-sample comparisons of level and power for simple goodness-of-fit statistics under cluster sampling. *Journal of the American Statistical Association* **82,** 630–636.

Titterington, D.M., and Bowman, A.W. (1985). A comparative study of smoothing procedures for ordered categorical data. *Journal of Statistical Computation and Simulation* **21,** 291–312.
Tukey, J.W. (1972). Some graphic and semigraphic displays. In *Statistical Papers in Honor of George W. Snedecor* (editor T.A. Bancroft), 293–316. Ames, IA, Iowa State University Press.
Tumanyan, S.Kh. (1956). Asymptotic distribution of the χ^2 criterion when the number of observations and number of groups increase simultaneously. *Theory of Probability and Its Applications* **1,** 117–131.
Uppuluri, V.R.R. (1969). Likelihood ratio test and Pearson's X^2 test in multinomial distributions. *Bulletin of the International Statistical Institute* **43,** 185–187.
Upton, G.J.G. (1978). *The Analysis of Cross-Tabulated Data.* New York, John Wiley.
Upton, G.J.G. (1982). A comparison of alternative tests for the 2×2 comparative trial. *Journal of the Royal Statistical Society Series A* **145,** 86–105.
van der Lubbe, J.C.A. (1986). An axiomatic theory of heterogeneity and homogeneity. *Metrika* **33,** 223–245.
Velleman, P.F., and Hoaglin, D.C. (1981). *Applications, Basics, and Computing of Exploratory Data Analysis.* Boston, Duxbury Press.
Verbeek, A., and Kroonenberg, P.M. (1985). A survey of algorithms for exact distributions of test statistics in $r \times c$ contingency tables with fixed margins. *Computational Statistics and Data Analysis* **3,** 159–185.
Wakimoto, K., Odaka, Y., and Kang, L. (1987). Testing the goodness of fit of the multinomial distribution based on graphical representation. *Computational Statistics and Data Analysis* **5,** 137–147.
Wald, A. (1943). Tests of statistical hypotheses concerning several parameters when the number of observations is large. *American Mathematical Society Transactions* **54,** 426–482.
Watson, G.S. (1957). The χ^2 goodness-of-fit test for normal distributions. *Biometrika* **44,** 336–348.
Watson, G.S. (1958). On chi-square goodness-of-fit tests for continuous distributions. *Journal of the Royal Statistical Association Series B* **20,** 44–61.
Watson, G.S. (1959). Some recent results in chi-square goodness-of-fit tests. *Biometrics* **15,** 440–468.
West, E.N., and Kempthorne, O. (1971). A comparison of the chi^2 and likelihood ratio tests for composite alternatives. *Journal of Statistical Computation and Simulation* **1,** 1–33.
Wieand, H.S. (1976). A condition under which the Pitman and Bahadur approaches to efficiency coincide. *Annals of Statistics* **4,** 1003–1011.
Wilcox, R.R. (1982). A comment on approximating the X^2 distribution in the equiprobable case. *Communications in Statistics—Simulation and Computation* **11,** 619–623.
Wilks, S.S. (1938). The large-sample distribution of the likelihood ratio for testing composite hypotheses. *Annals of Mathematical Statistics* **9,** 60–62.
Williams, C. (1950). On the choice of the number and width of classes for the chi-square test of goodness-of-fit. *Journal of the American Statistical Association* **45,** 77–86.
Williams, D.A. (1976). Improved likelihood ratio tests for complete contingency tables. *Biometrika* **63,** 33–37.
Wise, M.E. (1963). Multinomial probabilities and the χ^2 and X^2 distributions. *Biometrika* **50,** 145–154.
Yarnold, J.K. (1970). The minimum expectation in X^2 goodness of fit tests and the accuracy of approximations for the null distribution. *Journal of the American Statistical Association* **65,** 864–886.
Yarnold, J.K. (1972). Asymptotic approximations for the probability that a sum of

lattice random vectors lies in a convex set. *Annals of Mathematical Statistics* **43,** 1566–1580.

Yates, F. (1984). Tests of significance for 2 × 2 contingency tables. *Journal of the Royal Statistical Society Series A* **147,** 426–463.

Zelterman, D. (1984). Approximating the distribution of goodness of fit tests for discrete data. *Computational Statistics and Data Analysis* **2,** 207–214.

Zelterman, D. (1986). The log-likelihood ratio for sparse multinomial mixtures. *Statistics and Probability Letters* **4,** 95–99.

Zelterman, D. (1987). Goodness-of-fit tests for large sparse multinomial distributions. *Journal of the American Statistical Association* **82,** 624–629.

Author Index

Subject Index